Ministry of Agriculture Fisheries and Food

National Food Survey 1992

Annual Report on Household Food Consumption and Expenditure

LONDON : HMSO

© Crown Copyright 1993
Applications for reproduction
should be made to HMSO

First published 1993

ISBN 0 11 242955 6

Preface

The National Food Survey is the longest-running continuous survey of household food consumption and expenditure in the world. Since its inception in 1940, the Survey has provided a wealth of detailed information which has made a major contribution to the study of changing patterns of household food consumption.

This year's Report appears under a new title which better reflects its content. For the first time, information is included about purchases of alcoholic drinks and confectionery brought into the home. Results from the extension to cover details of food and drink consumed outside the home are being evaluated and may be presented in future Reports. A special section in this Report examines the consumption of fats (purchased as such) and the total fat content of the diet.

The Ministry of Agriculture, Fisheries and Food wishes to record its thanks to the many households who have provided information for the Survey. In addition, thanks are due to the staff of the Social Survey Division of the Office of Population Censuses and Surveys, who selected the samples, and of Social and Community Planning Research, who conducted the fieldwork and coding. They, and members of the advisory National Food Survey Committee, provided much help and methodological advice to the Ministry, which has overall responsibility for the Survey and for the processing of the results.

R E Mordue
(Chairman - National Food Survey Committee)

The National Food Survey Committee

R E MORDUE, BSc, MS
Ministry of Agriculture, Fisheries and Food, Chairman

J A BEAUMONT, BSc, PhD
Institute of Grocery Distribution

PROFESSOR A CHESHER, BSocSc
University of Bristol

W H B DENNER, BSc, PhD
Ministry of Agriculture, Fisheries and Food

P J LUND, BA, PhD
Ministry of Agriculture, Fisheries and Food

R U REDPATH, BA, MA
Office of Population Censuses and Surveys

PROFESSOR C RITSON, BA, MAgricSc
University of Newcastle upon Tyne

R G WHITEHEAD, BSc, PhD, MA, FIBiol
MRC Dunn Nutrition Group

M J WISEMAN, MB, MRCP
Department of Health

LESLEY YEOMANS, BSc, PhD
Tate and Lyle

Secretaries

D H BUSS, PhD, FRSH
Ministry of Agriculture, Fisheries and Food

SHEILA M DIXON, BSc
Ministry of Agriculture, Fisheries and Food

Contents

Page

List of Tables and Figures	ix
Section 1 Introduction	1
Section 2 Results of the 1992 Survey	
National Averages	4
Regional Comparisons	14
Income Group Comparisons	18
Analysis by Household Composition	23
Analysis by Household Composition and Income	27
Analysis by Age of Main Diary Keeper	30
Analysis by Ownership of Microwaves and Freezers	32
Section 3 Nutritional Results	
National Averages	33
Regional, Income Group and Household Composition Differences	36
Section 4 Special Analysis - Fats	
Cooking and Spreading Fats	39
Total Fat Intakes	49
Appendix A - Structure of the Survey	59
Appendix B - Supplementary Tables	67
Glossary and Additional Information	95

List of Tables and Figures

Page

1. Introduction

Table 1.1   Consumers' expenditure in the United Kingdom   2

Figure 1.2   Consumers' expenditure at current prices   2

Figure 1.3   Retail prices for food and for other items   2

2. Results of the 1992 Survey

Table 2.1   Household food expenditure and total value of food obtained for consumption   3

Figure 2.2   Changes in expenditure, prices and quantity of food in 1992, compared with 1991   3

Table 2.3   Consumption and expenditure for main food groups   4

Figure 2.4   Composition of expenditure on household food and drink, 1992   5

Table 2.5   Consumption and expenditure for milk and cheese   5

Table 2.6   Consumption and expenditure for meat, fish and eggs   6

Table 2.7   Consumption and expenditure for fats   7

Table 2.8   Consumption and expenditure for sugar and preserves   7

Table 2.9   Consumption and expenditure for vegetables and fruit   8

Table 2.10   Consumption and expenditure for bread, cereals and cereal products   9

Table 2.11   Consumption and expenditure for beverages and miscellaneous foods   10

Table 2.12   Purchases and expenditure for drinks and confectionery brought home   11

Table 2.13   Number of meals out (not from household supply)   12

Table 2.14   Family Expenditure Survey estimates of expenditure on food   12

Figure 2.15   Average number of mid-day meals per week per child aged 5 to 14 years   13

Table 2.16   Consumption and expenditure for selected foods by region, 1992   15

Figure 2.17   Consumption of fish by region, 1992   16

Figure 2.18   Consumption of cakes and biscuits by region, 1992   16

Figure 2.19   Consumption of alcoholic drinks brought home, by region, 1992   17

Table 2.20	Consumption and expenditure for selected foods by income group, 1992	18
Figure 2.21	Consumption and expenditure for carcase meat by income group, 1992	20
Figure 2.22	Consumption and expenditure for fish by income group, 1992	21
Figure 2.23	Consumption and expenditure for confectionery brought home, by income group, 1992	22
Table 2.24	Consumption of selected foods by household composition, 1992	23
Figure 2.25	Expenditure on main food groups per person by number of people in adult-only households, 1992	24
Figure 2.26	Expenditure on main food groups per person by number of children in 2-adult households, 1992	24
Figure 2.27	Expenditure on vegetables by household composition, 1992	25
Figure 2.28	Expenditure on soft drinks brought home and fruit juice by household composition, 1992	26
Figure 2.29	Consumption of confectionery brought home, by household composition, 1992	26
Figure 2.30	Total household food expenditure per head by certain household composition groups within income groups, 1992	27
Figure 2.31	Expenditure on selected foods by certain household composition groups within income groups, 1992	28
Table 2.32	Consumption and expenditure for selected foods by age of main diary-keeper, 1992	30
Table 2.33	Expenditure on selected foods by ownership of microwave or freezer, 1992	32

3. Nutritional Results

Table 3.1	Contributions made by groups of foods to household energy intakes in selected years	34
Table 3.2	Proportions of household food energy derived from fats and carbohydrate	35
Table 3.3	Intakes of vitamin C, β-carotene and folate in selected regions	36

4. Special Analysis - Fats

Figure 4.1	Consumption of fats, 1975 - 1992	39
Figure 4.2	Composition of consumption of yellow fats	40
Figure 4.3	Composition of consumption of other fats	41
Figure 4.4	Consumption of fats by income group	42
Figure 4.5	Relative consumption of butter by income group	43
Figure 4.6	Consumption of hard margarine by income group	43
Figure 4.7	Consumption of soft margarine by income group	44
Figure 4.8	Consumption of vegetable and salad oils by income group	44
Figure 4.9	Relative consumption of fats by household composition	45
Figure 4.10	Consumption of butter by household composition	46
Figure 4.11	Consumption of fats by age of main diary-keeper	47
Figure 4.12	Consumption of butter as percentage of total fats by age group of main diary-keeper	48
Figure 4.13	Energy and fat intakes, 1943 - 1992	50
Figure 4.14	Percentage of food energy from carbohydrate, fat and protein, 1943 - 1992	50
Table 4.15	Intakes of saturated, monounsaturated and polyunsaturated fatty acids	51
Figure 4.16	Contributions of foods to fat intakes	52
Figure 4.17	Contributions of foods to saturated, monounsaturated, and polyunsaturated fatty acid intakes	53
Figure 4.18	Trends in proportions of food energy as fat, carbohydrate and saturated fatty acids by income group	55

Section 1

Introduction

The National Food Survey data presented in this Report were derived from the responses of a random sample of private households throughout Great Britain. Each of the 7,556 participating households recorded details of all items of food for human consumption brought into the home during the course of a week. Information on soft and alcoholic drinks and confectionery brought home was also included. Some information on the numbers of meals eaten outside the home was recorded for all households, but not the content or cost of such meals. In addition, members of selected households recorded details of all meals, snacks and drinks consumed outside the home; the results of this recording are still being evaluated.

The main results for 1992 are presented in Section 2 of the Report. They show consumption and expenditure data for major types of food expressed as averages per person per week. Aggregate data for Great Britain as a whole are followed by analyses by various household characteristics. These provide valuable insights into patterns of consumption and expenditure in different types of household, but need to be interpreted with some care as the differences observed cannot necessarily be attributed solely to the classification difference under consideration. For example, differences in levels of expenditure between income groups may partly reflect differences in the numbers and ages of household members and the number of meals eaten outside the home. Information on purchases of soft and alcoholic drinks and confectionery brought home is also presented in this Section. Section 3 presents the nutritional data for 1992. Section 4 presents a special analysis of consumption of fats, analysing firstly changes in consumption of cooking and spreading fats over the period from 1975 to 1992, and secondly changes in intakes and sources of all forms of fat in the diet from a nutritional viewpoint. Appendix A contains details of the structure and conduct of the Survey and Appendix B contains supplementary tables showing the main results for 1992.

An estimated £45 billion was spent on household food in 1992 (Table 1.1), just under 12 per cent of total consumer expenditure. This proportion has declined considerably since 1950 due to the much greater increase in consumers' expenditure on other items over the same period (Figure 1.2).

Table 1.1
Consumers' expenditure in the United Kingdom

	1982		1987		1992	
	£ b	%	£ b	%	£ b	%
Expenditure on household food	26.5	15.6	34.4	13.0	45.3	11.8
Total consumers' expenditure	169.4	100.0	265.3	100.0	382.7	100.0
Related series:						
Expenditure on alcoholic drink	12.0	7.1	17.5	6.6	24.6	6.4
Expenditure on catering	9.5	5.6	18.1	6.8	32.8	8.6

Source: Central Statistical Office

Figure 1.2
Consumers' expenditure at current prices

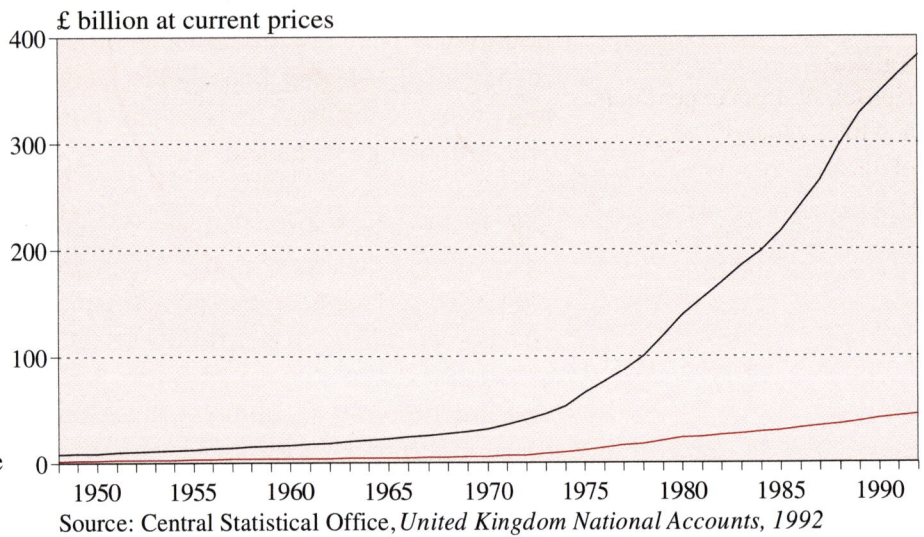

Source: Central Statistical Office, *United Kingdom National Accounts, 1992*

The relatively slow growth in expenditure on household food is due in part to the smaller increase in retail food prices compared to other items, particularly over the last 20 years (Figure 1.3). It will also have been affected by increasing expenditure on eating out over much of the same period. Nevertheless household expenditure on food at constant prices has risen by a third since 1950.

Figure 1.3
Retail prices for food and for other items

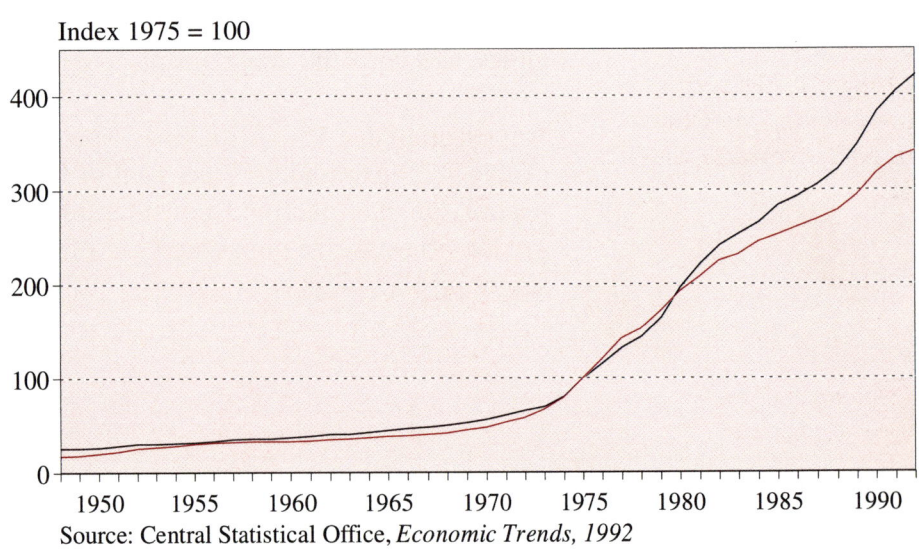

Source: Central Statistical Office, *Economic Trends, 1992*

Section 2

Results of the 1992 Survey

Average expenditure on household food in each quarter of 1992 was higher than in each of the corresponding quarters of the previous year, resulting in a 1.9 per cent annual increase over 1991 (Table 2.1). The total value of food consumption was estimated at £13.09 per person per week compared with £12.89 in 1991. Both figures include the value of garden and allotment produce, which declined further in 1992. A further £1.58 was spent on soft and alcoholic drinks and confectionery.

Table 2.1
Household food expenditure and total value of food obtained for consumption

per person per week

	Expenditure on food			Value of garden and allotment produce, etc[a]		Value of consumption[b]		
	1991	1992	Change	1991	1992	1991	1992	Change
	£	£	%	£	£	£	£	%
1st quarter	12.18	12.45	+2.2	0.11	0.12	12.29	12.57	+2.3
2nd quarter	13.18	13.37	+1.4	0.13	0.11	13.31	13.48	+1.3
3rd quarter	12.64	12.86	+1.7	0.34	0.26	12.98	13.12	+1.1
4th quarter	12.77	13.04	+2.1	0.19	0.14	12.96	13.18	+1.7
Yearly average	**12.69**	**12.93**	**+1.9**	**0.20**	**0.16**	**12.89**	**13.09**	**+1.6**
Drinks and sweets[c]	na	1.58	–	na	–	na	1.58	–
Total food and drink	**na**	**14.51**	**–**	**na**	**0.17**	**na**	**14.67**	**–**

(a) valued at average prices paid for comparable purchases

(b) expenditure on food purchased for consumption in the home, plus the value of garden and allotment produce, etc

(c) includes soft drinks, confectionery and alcoholic drinks

Prices paid for household food in 1992 averaged 1.6 per cent more than in the previous year (Figure 2.2). The estimated quantity of food purchased rose by 0.2 per cent. Within this total, convenience food

Figure 2.2
Changes in expenditure, prices and quantity of food in 1992, compared with 1991

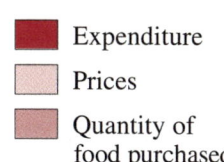
■ Expenditure
▢ Prices
▨ Quantity of food purchased

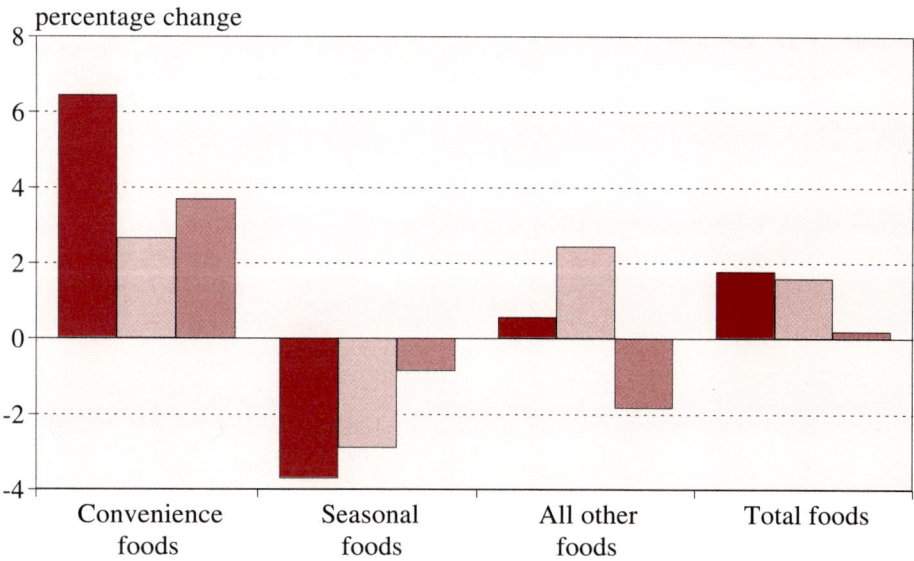

showed the largest increase in average price (2.7 per cent) while seasonal food prices fell. Further details of the average prices paid for individual foods are given in Appendix Table B2.[1]

National Averages

In this sub-section the national average results for 1992 are presented and compared with those for 1990 and 1991. The 1992 results indicate that household expenditure on most major food groups has risen since 1991 (Table 2.3) but there are some notable exceptions. Household expenditure on eggs, sugar and preserves, fruit, and beverages was lower than in 1991 and there was no change in total expenditure on meat and meat products. Milk and cream and also cereal products showed an increase in consumption compared with 1991. Expenditure on cheese rose by 11 per cent since 1990 although its consumption remained steady at 4 ounces per person per week. Fish consumption has remained consistently around 5 ounces per person per week since 1985. Eggs and sugar and preserves show the most marked drop in consumption, of about 7½ per cent on 1991. Consumption of vegetables has continued to decline but expenditure has risen due to the purchase of more highly prepared items. Expenditure on soft drinks, confectionery and alcoholic drinks constituted about 11 per cent of total expenditure in 1992 (Figure 2.4).

Table 2.3
Consumption and expenditure for main food groups

per person per week

		Consumption			Expenditure		
		1990	1991	1992	1990	1991	1992
		(ounces)[a]			(pence)		
Milk and cream	(pt or eq pt)	3.82	3.74	3.91	127.69	134.45	143.53
Cheese		4.00	4.11	4.01	41.43	43.45	46.03
Meat and meat products		34.11	33.93	33.52	337.24	339.73	339.86
Fish		5.08	4.90	4.99	66.62	66.58	68.32
Eggs	(no)	2.20	2.25	2.08	19.49	20.13	18.77
Fats and oils		9.00	8.76	8.66	36.11	36.41	37.40
Sugar and preserves		7.73	7.67	7.10	18.08	19.58	18.27
Vegetables		79.79	78.42	77.36	169.45	180.97	183.75
Fruit		31.56	33.56	32.79	97.13	107.84	104.26
Cereals (incl. bread)		51.79	51.30	51.57	203.98	215.22	225.75
Beverages		2.47	2.50	2.35	43.18	45.90	43.11
Other foods		–	–	–	51.66	59.19	63.98
Total food		–	–	–	£12.12	£12.69	£12.93
Soft drinks	(fl oz)	21.42	22.92	25.39	32.60	36.89	43.24
Alcoholic drinks	(cl)	na	na	30.78	na	na	90.98
Confectionery	(g)	na	na	50.92	na	na	23.61
Total all food and drink		–	–	–	–	–	**£14.51**

(a) except where otherwise stated

[1] It should be noted that since the results for household consumption presented in this Report include both purchases and 'free food', average prices paid for purchases cannot in general be derived by dividing the expenditure on a particular food by average consumption.

Figure 2.4
Composition of expenditure on household food and drink, 1992

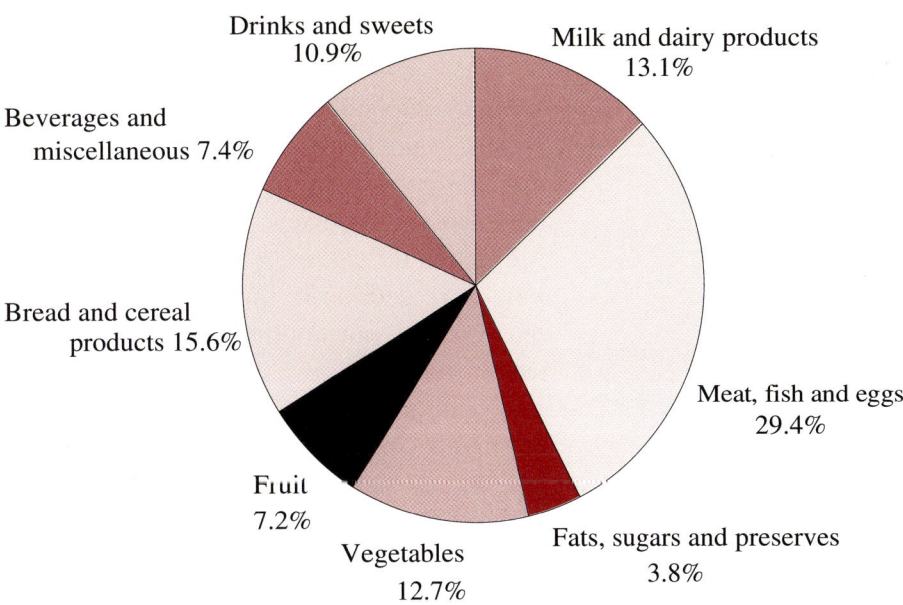

Milk, cream and cheese

Household consumption of milk and cream rose to 3.91 pints per person per week during 1992, of which 3.45 pints were obtained as liquid milk (Table 2.5). Most of the increase was due to higher consumption of low fat milk which, at 1.7 pints per person per week, equalled that of liquid whole milk for the first time. Yoghurt and fromage frais consumption continued to rise, while cheese consumption was at a similar level to that in 1990 but with a higher proportion of processed cheese.

Table 2.5
Consumption and expenditure for milk and cheese

per person per week

		Consumption			Expenditure		
		1990	1991	1992	1990	1991	1992
		(ounces)[a]			(pence)		
MILK AND CREAM:							
Liquid wholemilk, full price	(pt)	2.12	1.90	1.69	61.80	58.35	53.23
Welfare and school milk	(pt)	0.04	0.04	0.06	0.12	0.13	0.21
Low fat milks	(pt)	1.25	1.37	1.70	36.23	41.55	52.52
Dried and other milk [b]	(pt or eq pt)	0.21	0.21	0.22	8.99	10.30	10.15
Yoghurt and fromage frais	(pt)	0.17	0.19	0.21	16.55	19.79	23.19
Cream	(pt)	0.02	0.03	0.03	4.00	4.33	4.25
Total milk and cream		**3.82**	**3.74**	**3.91**	**127.69**	**134.45**	**143.53**
CHEESE:							
Natural		3.70	3.78	3.66	37.72	39.31	41.50
Processed		0.30	0.33	0.35	3.71	4.14	4.52
Total cheese		**4.00**	**4.11**	**4.01**	**41.43**	**43.45**	**46.03**

(a) except where otherwise stated.
(b) including condensed milk

Meat, fish and eggs

Average expenditure on meat was £3.40 per person per week in 1992, which is unchanged from 1991 but with less spent on carcase meat and more on meat products (Table 2.6). Carcase meat consumption declined to 10 ounces per person per week with the greatest reduction, for lamb, reflecting a marked increase in average prices paid. The long term decline in purchases of uncooked bacon and ham continued, but poultry consumption increased sharply to the highest annual value so far recorded.

Although overall fish consumption has been fairly constant over the last decade the purchase of fresh fish has gradually declined since the early 1980s to just over one ounce per person per week in 1992. Over the same period, consumption of prepared fish has risen as a proportion of the total. Egg consumption fell sharply to 2.08 eggs per person per week.

Table 2.6
Consumption and expenditure for meat, fish and eggs

per person per week

	Consumption 1990	Consumption 1991	Consumption 1992	Expenditure 1990	Expenditure 1991	Expenditure 1992
	(ounces)[a]			(pence)		
MEAT:						
Beef and veal	5.24	5.35	4.98	63.89	67.77	62.40
Mutton and lamb	2.92	3.02	2.49	28.64	29.17	26.54
Pork	2.97	2.88	2.53	29.07	27.62	24.75
Total carcase meat	**11.12**	**11.25**	**10.00**	**121.60**	**124.56**	**113.69**
Bacon and ham, uncooked	3.02	3.00	2.73	33.92	33.08	32.05
Poultry, uncooked	7.43	7.14	7.64	50.02	48.48	50.01
Other meat and meat products	12.55	12.54	13.15	131.71	133.61	144.11
Total meat	**34.11**	**33.93**	**33.52**	**337.24**	**339.73**	**339.86**
FISH:						
Fresh	1.14	1.15	1.02	15.04	16.31	15.25
Processed and shell	0.51	0.56	0.55	8.66	10.13	10.37
Prepared, including fish products	1.79	1.64	1.92	24.93	21.80	25.14
Frozen, including fish products	1.65	1.53	1.50	17.99	18.34	17.57
Total fish	**5.08**	**4.90**	**4.99**	**66.62**	**66.58**	**68.32**
EGGS (no)	**2.20**	**2.25**	**2.08**	**19.49**	**20.13**	**18.77**

(a) except where otherwise stated

Fats and oils

Total purchases of fats were one per cent lower than in 1991 (Table 2.7). Consumption of low fat and dairy spreads and of vegetable and salad oils continued to increase strongly while butter and margarine declined further. Expenditure on fats rose by 2.7 per cent compared with 1991 in spite of the drop in consumption.

Table 2.7
Consumption and expenditure for fats

per person per week

	Consumption			Expenditure		
	1990	1991	1992	1990	1991	1992
	(ounces)			(pence)		
FATS:						
Butter	1.61	1.54	1.44	11.01	10.43	10.02
Margarine	3.19	3.14	2.79	9.97	10.79	10.05
Low fat and dairy spreads	1.58	1.66	1.80	7.89	8.16	9.12
Vegetable and salad oils	1.53	1.52	1.73	3.77	3.92	4.88
Other fats and oils	1.08	0.89	0.88	3.46	3.11	3.33
Total fats	**9.00**	**8.76**	**8.66**	**36.11**	**36.41**	**37.40**

Sugar and preserves

Consumption of sugar and preserves fell back sharply during 1992 (Table 2.8). Sugar consumption, at 5½ ounces per person per week, continued its strong downward trend. Despite a recovery in 1991, purchases of preserves fell and consumption was 6 per cent lower than in 1990.

Table 2.8
Consumption and expenditure for sugar and preserves

per person per week

	Consumption			Expenditure		
	1990	1991	1992	1990	1991	1992
	(ounces)			(pence)		
SUGAR AND PRESERVES:						
Sugar	6.04	5.88	5.51	11.10	11.62	10.61
Honey, preserves, syrup and treacle	1.69	1.79	1.59	6.98	7.96	7.66
Total sugar and preserves	**7.73**	**7.67**	**7.10**	**18.08**	**19.58**	**18.27**

Vegetables and fruit

Total consumption of vegetables was a little lower in 1992 than in the previous year, reflecting the continuing decline in consumption of potatoes (Table 2.9). Amounts of fresh green vegetables, particularly cabbages and beans, were lower, although there was some increase for cauliflowers. These reductions were offset, however, by higher consumption of other fresh vegetables. Overall purchases of frozen produce in 1992 were similar to 1991, but there was a continued trend towards more frozen convenience potato products and less peas and beans. Spending on chips and other potato products also rose significantly.

Table 2.9
Consumption and expenditure for vegetables and fruit

per person per week

	Consumption			Expenditure		
	1990	1991	1992	1990	1991	1992
	(ounces)[a]			(pence)		
VEGETABLES:						
Potatoes	35.17	33.81	31.78	27.53	28.05	25.12
Fresh green	9.79	9.13	8.81	21.95	24.18	22.95
Other fresh	16.19	16.26	16.74	46.27	49.89	49.01
Frozen, including vegetable products	6.54	7.04	7.00	20.88	23.68	23.56
Other processed, including vegetable products	12.10	12.19	13.03	52.81	55.17	63.11
Total vegetables	**79.79**	**78.43**	**77.36**	**169.45**	**180.97**	**183.75**
FRUIT:						
Fresh	21.33	21.53	21.80	65.06	70.22	69.28
Fruit juices (fl oz)	7.11	8.80	7.81	17.25	21.17	18.26
Other fruit products	3.13	3.23	3.18	14.83	16.45	16.72
Total fruit	**31.56**	**33.56**	**32.79**	**97.13**	**107.84**	**104.26**

(a) except where otherwise stated

Consumption of fresh fruit rose again slightly in 1992, partly due to increases in purchases of bananas and grapes although amounts of oranges and apples were less. Expenditure, however, was lower as some prices which had risen steeply in 1991 - particularly the average price paid for apples - declined somewhat. Consumption of fruit juices and other fruit products fell back a little.

Bread, cereals and cereal products

Bread consumption showed little change in 1992, in contrast to the decline in recent years, largely due to a rise in the purchase of wholemeal and premium white loaves, and Vienna and French bread (Table 2.10) for which higher prices are paid. This offset a continued decline in purchases of standard white loaves. Overall expenditure on bread rose by one per cent.

Table 2.10
Consumption and expenditure for bread, cereals and cereal products

per person per week

	Consumption			Expenditure		
	1990	1991	1992	1990	1991	1992
	(ounces)			(pence)		
BREAD:						
White bread (standard loaves)	14.76	13.71	12.95	28.37	27.27	24.46
White premium and softgrain bread[a]	na	na	2.20	na	na	4.70
Brown bread	3.45	3.64	3.29	8.30	9.17	8.32
Wholemeal bread	3.82	3.67	3.86	8.95	9.25	9.39
Other bread (incl. rolls and prepared sandwiches)	6.04	5.51	4.33	22.98	22.94	22.52
Total bread	**28.09**	**26.53**	**26.62**	**68.59**	**68.64**	**69.40**
OTHER CEREALS AND CEREAL PRODUCTS:						
Flour	3.19	2.84	2.84	3.55	3.54	3.74
Cakes and pastries	3.65	4.15	4.08	28.98	30.16	31.18
Biscuits	5.26	5.18	5.23	31.75	32.62	34.04
Oatmeal and oat products	0.52	0.69	0.53	2.21	2.81	1.66
Breakfast cereals	4.47	4.72	4.66	26.52	29.93	31.63
Other cereals	6.62	7.20	7.61	42.38	47.52	54.11
Total cereals, including bread	**51.79**	**51.30**	**51.57**	**203.98**	**215.22**	**225.75**

(a) largely included in 'other bread' before 1992

Flour consumption remained the same as in 1991 while purchases of breakfast cereals and cakes and pastries went down slightly. Consumption of certain cereal convenience foods, such as frozen cakes and pastries and canned pasta, rose and contributed to an increase in expenditure on all cereal products of nearly 5 per cent.

Beverages and miscellaneous foods

In 1992 consumption of tea was nearly ten per cent below the 1990 level (Table 2.11) and that of coffee reverted to its 1990 level. Expenditure on both tea and coffee was lower than in 1990, but a doubling of expenditure on branded food drinks since 1990 meant that total expenditure on beverages was largely unchanged.

Table 2.11
Consumption and expenditure for beverages and miscellaneous foods

per person per week

	Consumption			Expenditure		
	1990	1991	1992	1990	1991	1992
	(ounces)[a]			(pence)		
BEVERAGES:						
Tea	1.52	1.48	1.37	20.09	21.60	19.42
Coffee	0.63	0.67	0.63	19.82	20.34	19.36
Cocoa and drinking chocolate	0.16	0.14	0.12	1.69	1.34	1.14
Branded food drinks	0.16	0.21	0.23	1.58	2.62	3.19
Total beverages	**2.47**	**2.50**	**2.35**	**43.18**	**45.90**	**43.11**
MISCELLANEOUS:						
Mineral water[b] (fl oz)	1.96	2.45	3.43	2.10	2.89	4.23
Soups, canned, dehydrated and powdered	2.52	2.59	2.58	7.89	9.12	9.39
Pickles and sauces	2.37	2.44	2.55	11.32	12.95	14.65
Ice-cream, mousse (fl oz)	3.31	3.33	3.31	11.76	13.37	13.11
Other foods [c]	na	na	na	18.61	20.86	22.59
Total miscellaneous	**na**	**na**	**na**	**51.66**	**59.19**	**63.98**

(a) except where otherwise stated
(b) the 1992 results for mineral water are inflated by a single large purchase in a Survey household
(c) including spreads, salt and other miscellaneous food items

Drinks and confectionery brought home

Soft drinks have been included in the main analysis of the Survey results for the first time (Table 2.12). However, as for all the results presented, the estimates refer only to household purchases and exclude casual purchases not taken home or not brought to the attention of the main diary-keeper. Consumption of soft drinks has continued to rise steadily, the most noticeable trend being the increase in quantities of low calorie products.

Table 2.12

Purchases and expenditure for drinks and confectionery brought home

per person per week

	Purchase Quantity			Expenditure		
	1990	1991	1992	1990	1991	1992
	(fluid ounces)			(pence)		
SOFT DRINKS:[a]						
Concentrated	4.07	4.18	4.45	8.97	9.81	11.79
Unconcentrated	12.82	13.25	14.04	17.58	19.14	20.95
Low calorie concentrated	na	na	0.61	na	na	1.43
Low calorie unconcentrated[b]	4.55	5.50	6.26	6.12	7.94	9.06
All soft drinks[c]	**37.72**	**39.64**	**45.60**	**32.67**	**36.89**	**43.24**
	(centilitres)			(pence)		
ALCOHOLIC DRINKS:						
Lager and beer [d]	–	–	16.68	–	–	24.80
Wine	–	–	8.69	–	–	33.52
Other	–	–	5.27	–	–	32.66
Total alcoholic drinks	–	–	**30.64**	–	–	**90.98**
	(grams)			(pence)		
CONFECTIONERY:						
Chocolate confectionery	–	–	34.98	–	–	17.19
Mints and boiled sweets	–	–	13.79	–	–	5.46
Other	–	–	2.14	–	–	0.96
Total confectionery	–	–	**50.90**	–	–	**23.61**

(a) excluding pure fruit juices which are recorded in the Survey under fruit products
(b) includes some concentrated low calorie soft drinks before 1992
(c) converted to unconcentrated equivalent
(d) includes low alcohol lager and beer

Information on alcoholic drinks bought for consumption at home was collected by the Survey for the first time in 1992.[2] Just over half the volume of alcoholic drink consumed at home was lager and beer. This represents only just over a quarter of the total expenditure, which averaged 91 pence per person per week. The highest expenditure was on wine but spirits and liqueurs, included in the table with other drinks such as cider, fortified wines and ready mixed drinks, accounted for nearly a quarter of the total.

Confectionery brought into the home was also included in the Survey for the first time in 1992. Average expenditure was 24 pence per person per week, with consumption just over 50g per person. Over two-thirds of such confectionery consumption consisted of solid chocolate or chocolate bars and filled chocolates. Mints and boiled sweets took most of the remaining share while other sweets, including fudges, toffees, caramels and chewing gum, accounted for less than 5 per cent.

---

2 The consumption and expenditure estimates are expressed per person per week, in line with the other results of the Survey, even though not all members of the household (particularly the children) may consume these items.

Meals eaten outside the home

Information is recorded within the Survey on the number of meals bought and eaten outside the home (Table 2.13). In 1992 the number of mid-day meals eaten out (ie not from household supplies) declined to 1.72 per person per week.

Table 2.13
Number of meals out (not from household supply)

	1990	1991	1992
Mid-day meals out	1.93	1.84	1.72
All meals out[a]	na	na	2.78

per person per week

(a) based on a pattern of three meals a day

This change may be part of a continuing trend. The Family Expenditure Survey shows that expenditure on meals bought away from the home rose less steeply between 1987 and 1992 than between 1982 and 1987 (Table 2.14).

Table 2.14
Family Expenditure Survey estimates of expenditure on food

	1982	1987	1991	1992
Household food[a]	23.94	28.58	36.74	38.14
Meals bought away from home	4.25	7.21	9.39	9.52

£ per household per week

(a) including soft drinks, chocolate and sugar confectionery
Source: Family Expenditure Survey

The National Food Survey has collected information on the source of mid-day meals for children aged 5 to 14 continuously since 1972. The results (Figure 2.15) show that school meals have declined to an average of fewer than 2 mid-day meals per week whereas packed lunches and other meals out have increased steadily over the last twenty years.

Figure 2.15
Average number of mid-day meals per week per child aged 5 to 14 years

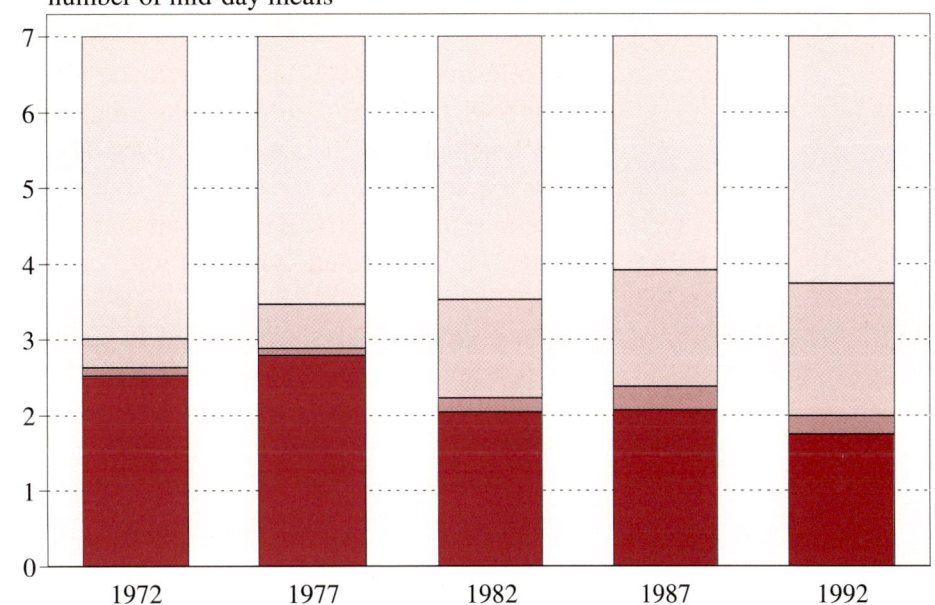

Regional Comparisons

While the Survey is designed to be representative of Great Britain as a whole, practical considerations limit the number of separate areas from each region which can be included in any one year (see Appendix A for the structure of the Survey). For this reason, comparisons between regions, and in particular comparisons between years for the same region, must be interpreted with caution.

The regional analysis presented in Table 2.16 shows that total weekly expenditure on household food ranged from £11.66 per person per week in the West Midlands region to £13.94 in the South East and East Anglia region. There was also considerable regional variation in the consumption of different types of food, with cheese, fish, fruit and cereal products other than bread showing the most variation.

Expenditure on meat and meat products was a third higher in Scotland than in the South West, although a slightly lower proportion of this was spent on carcase meat. More fruit and cheese was eaten in the South East and East Anglia than in the North of England. Bread consumption was highest in the West Midlands although, on average, consumers there spent least on it.

Although consumers in Scotland purchased less beverages than in the other regions, household consumption and expenditure on soft drinks was highest in Scotland, where ready-to-drink products had the largest share. In particular consumption was 50 per cent greater there than in the South West, where households bought a higher proportion of soft drinks as concentrates.

Fish consumption was highest in Yorkshire and Humberside with prepared items such as cooked or canned fish accounting for 65 per cent of the overall total (Figure 2.17). Households in Scotland had the highest average consumption of fresh or frozen white fish while the lowest amounts were consumed in Wales, the South West and West Midlands. Consumption of fat fish was highest in the North West, South East and South West.

Table 2.16
Consumption and expenditure for selected foods by region, 1992

per person per week

	North	Yorkshire and Humberside	North West	East Midlands	West Midlands	South West	South East/ East Anglia	England	Wales	Scotland	All households
CONSUMPTION					(ounces)[a]						
Milk and cream (pt or eq pt)	3.80	3.76	3.80	4.19	3.72	4.07	3.89	3.88	4.18	3.98	3.91
Cheese	3.16	3.79	3.57	4.01	4.35	4.14	4.31	4.04	3.77	3.83	4.01
Carcase meat	9.97	10.52	9.80	10.03	10.60	9.13	10.28	10.11	9.38	9.28	10.00
Other meat and meat products	26.60	24.07	26.10	20.84	23.47	21.15	22.76	23.40	24.74	23.85	23.52
Fish	4.97	6.03	5.11	4.95	4.23	4.27	5.27	5.03	4.86	4.69	4.99
Eggs (no)	2.32	2.20	2.00	2.00	2.05	2.12	2.00	2.06	2.15	2.28	2.08
Fats and oils	8.40	9.34	8.67	9.71	7.99	8.86	8.62	8.71	8.85	8.03	8.66
Sugar and preserves	6.39	7.65	6.93	7.82	7.68	7.70	6.67	7.09	7.86	6.70	7.10
Vegetables	81.55	79.95	75.69	79.46	76.44	82.40	77.05	78.09	79.62	68.62	77.36
Fruit	26.94	31.70	26.60	31.74	26.63	33.60	38.68	32.96	29.23	33.57	32.79
Bread	28.41	26.60	27.77	28.22	30.08	25.43	24.27	26.41	28.41	27.42	26.62
Other cereals	24.59	28.60	23.76	27.59	20.31	25.31	25.67	25.03	24.26	24.54	24.94
Beverages	2.06	2.41	2.49	2.42	2.13	2.48	2.46	2.39	2.46	1.87	2.35
Soft drinks (fl oz)	24.16	26.29	26.39	23.59	24.84	20.40	25.57	24.87	25.07	30.72	25.39
Alcoholic drinks (cl)	32.19	26.69	36.77	24.42	27.87	28.14	34.37	31.70	24.80	25.77	30.78
Confectionery (g)	51.47	49.14	43.06	53.91	49.34	54.65	51.98	50.52	43.93	59.72	50.92
EXPENDITURE					(pence)						
Milk and cream	133.03	137.58	132.48	146.01	131.05	147.53	153.18	143.59	146.02	141.17	143.53
Cheese	34.34	41.87	37.97	44.14	48.23	48.92	52.23	46.52	39.82	45.45	46.03
Carcase meat	106.33	119.84	115.80	107.51	114.73	100.84	116.39	113.36	101.69	125.37	113.69
Other meat and meat products	225.47	227.28	244.81	185.54	210.45	191.66	231.45	222.32	224.93	265.23	226.17
Fish	59.29	78.20	68.17	63.68	57.53	54.81	76.14	68.51	63.00	70.22	68.32
Eggs	19.53	18.30	16.61	16.50	17.77	18.73	19.49	18.45	20.36	20.85	18.77
Fats and oils	32.77	38.24	35.46	38.13	33.76	38.64	39.50	37.43	38.54	36.36	37.40
Sugar and preserves	16.42	19.36	17.18	18.87	16.90	21.18	18.23	18.23	18.64	18.36	18.27
Vegetables	168.67	176.69	172.00	163.34	167.94	172.21	205.54	184.25	180.62	180.90	183.75
Fruit	79.24	98.31	83.50	93.66	79.91	98.58	128.65	104.21	91.21	113.84	104.26
Bread	74.13	69.67	70.63	67.00	66.08	66.75	67.46	68.32	72.53	77.93	69.40
Other cereals	153.15	167.77	151.27	145.96	130.47	150.45	165.82	155.51	150.92	168.42	156.35
Beverages	37.61	44.38	44.75	44.05	38.81	45.47	44.74	43.50	43.91	38.79	43.11
Other foods	60.37	59.97	58.92	52.28	53.18	58.07	75.34	64.29	54.91	67.17	63.98
Total food	**£12.00**	**£12.97**	**£12.50**	**£11.87**	**£11.67**	**£12.14**	**£13.94**	**£12.88**	**£12.47**	**£13.70**	**£12.93**
Soft drinks	35.89	40.02	43.57	37.07	37.85	35.88	47.12	42.12	41.45	55.51	43.24
Alcoholic drinks	68.43	66.80	93.42	64.89	75.90	84.33	111.14	91.05	76.35	100.41	90.98
Confectionery	23.88	22.30	19.40	23.79	22.29	25.65	24.67	23.39	19.77	28.39	23.61
Total all food and drink	**£13.29**	**£14.27**	**£14.06**	**£13.12**	**£13.03**	**£13.60**	**£15.77**	**£14.45**	**£13.85**	**£15.54**	**£14.51**

(a) except where otherwise stated

Figure 2.17
Consumption of fish by region, 1992

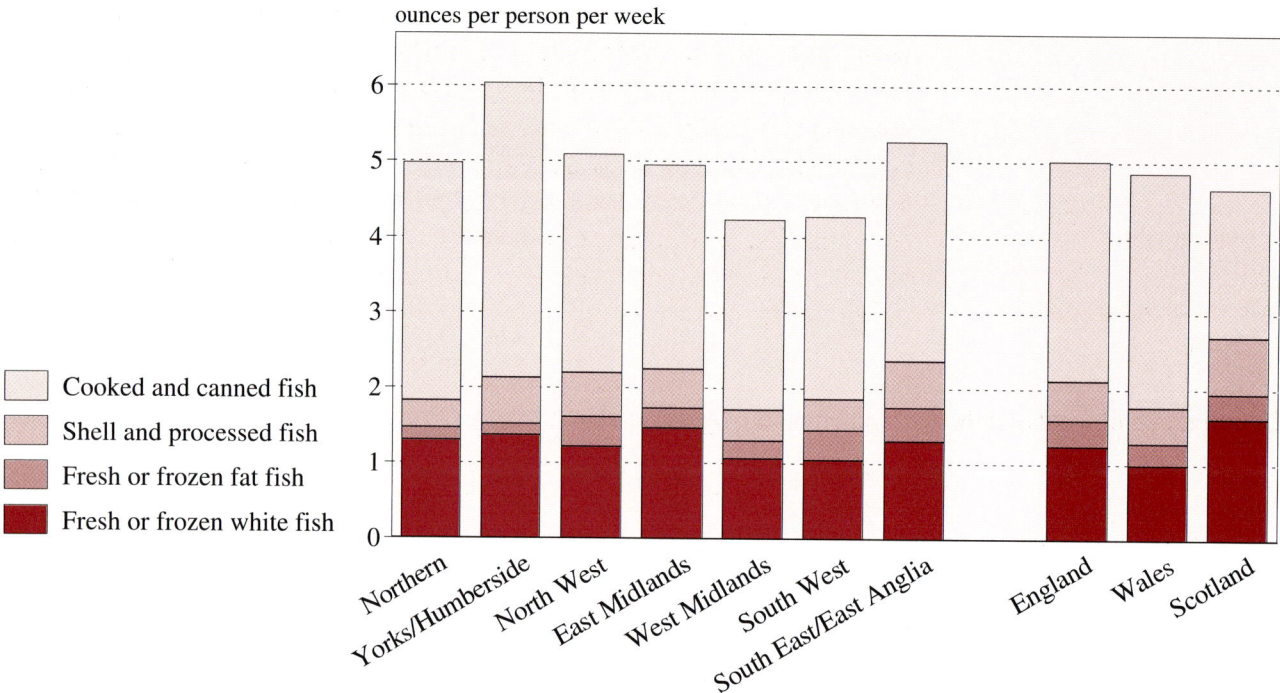

Cake and biscuit purchases were higher in the Northern regions of England than in the Midlands and the South, with consumption of buns, scones and teacakes particularly high in Yorkshire and Humberside and the North West. The South and Midlands regions had similar overall consumption of cakes and buns, but the South West households bought more biscuits.

Figure 2.18
Consumption of cakes and biscuits by region, 1992

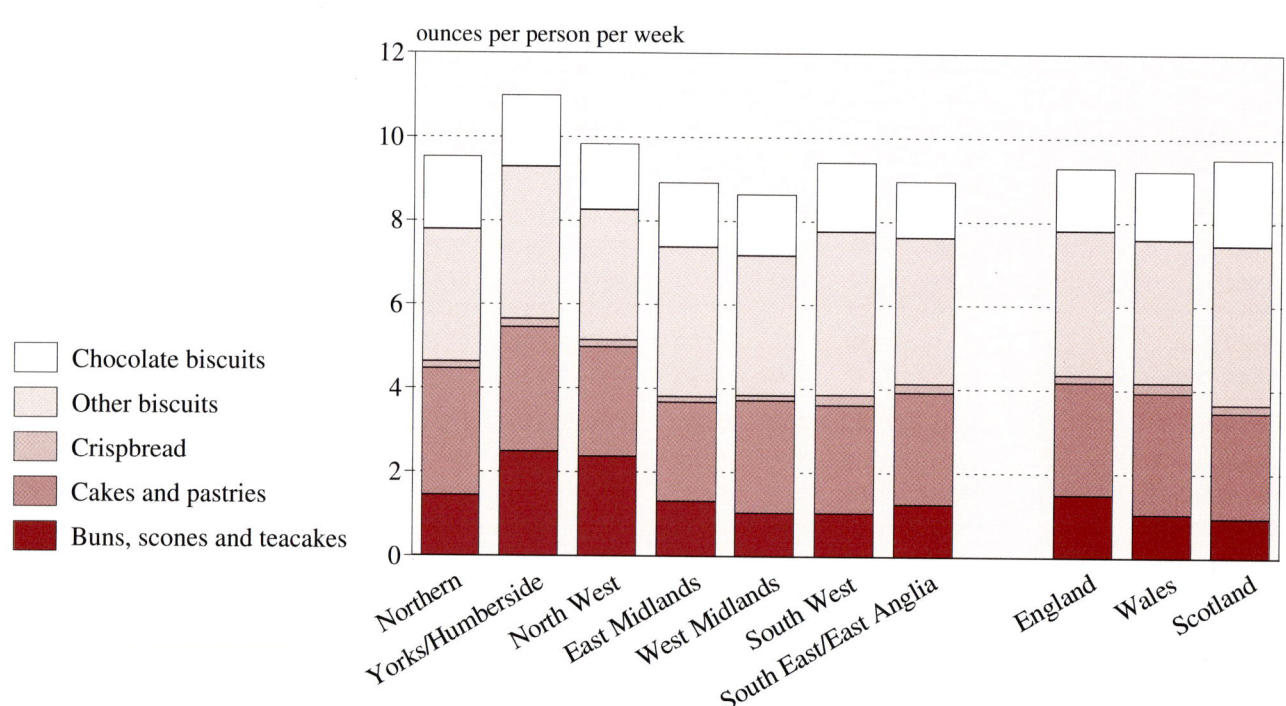

Expenditure on alcoholic drinks for the home was greatest in the South East and East Anglia and in Scotland. The highest consumption by volume occurred in the North, North West and South East of England, with lager the most prevalent in the northern regions and wine in the South East (Figure 2.19). In Scotland spirits and liqueur consumption was much higher than in any other region. Beer (excluding lager) comprised about a third of the alcoholic drinks consumed by West Midlands households, the highest proportion in any region. Cider consumption was greatest in Welsh households but in Wales, together with the East Midlands region, overall consumption of alcoholic drinks in the home was lowest.

Figure 2.19
Consumption of alcoholic drinks brought home, by region, 1992

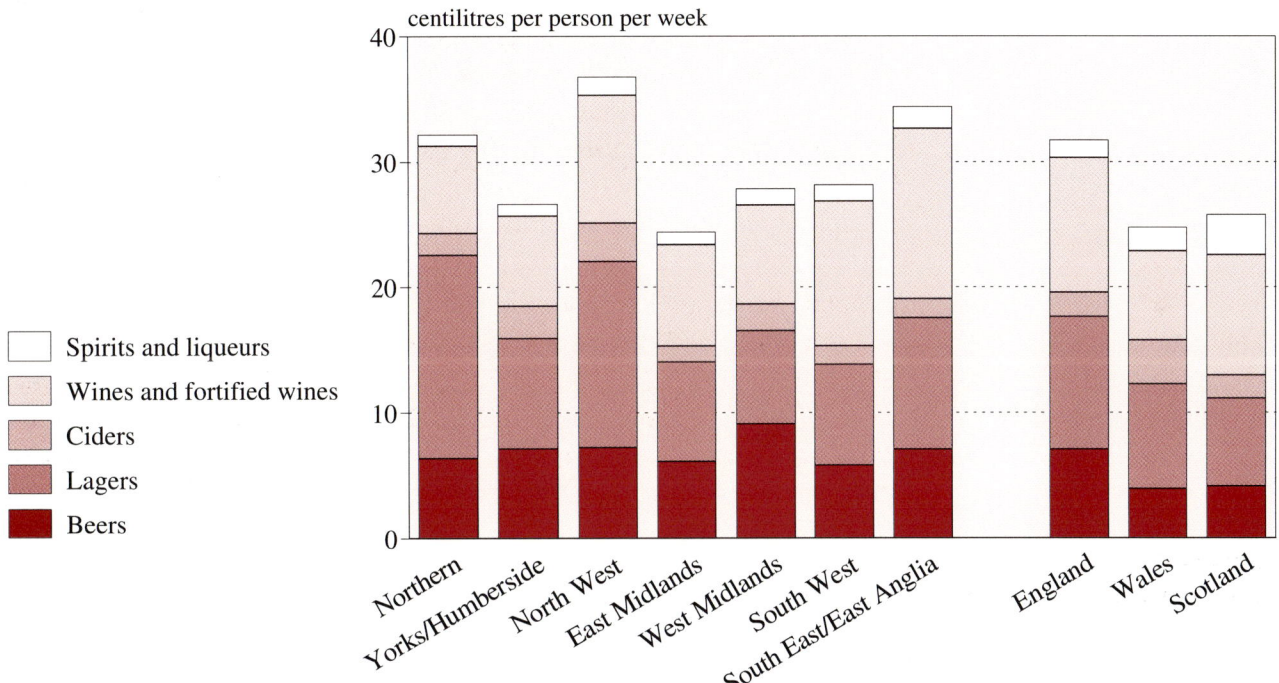

Income Group Comparisons

The distribution of households by income group[3] always differs slightly from the target distribution and from that achieved in previous years. As a result, the estimates of food consumption and expenditure analysed by this characteristic (Table 2.20) will not be entirely comparable with those of earlier years, but they do reveal some patterns in household food purchasing between households of different incomes.

For those households with at least one earner, average expenditure per person on household food increased, with the income of the head of the household, from £11.10 per week in income group D to £15.39 in income group A. The highest average total expenditure, £16.42, occurred in the non-earner group E1. The expenditure figures for the pensioner and non-earner categories are higher than for earning households at a comparable income level; this reflects a number of factors, for example households in these groups typically contained fewer children and had more meals at home.

Per capita spending on most of the component food groups also increased with income; the exception is eggs, for which expenditure declined with

Table 2.20
Consumption and expenditure for selected foods by income group, 1992

per person per week

INCOME GROUP[a]

Gross weekly income of head of household

		Households with one or more earners				Households without an earner		OAP
		A	B	C	D	E1	E2	
CONSUMPTION				(ounces)[b]				
Milk and cream	(pt or eq pt)	3.87	3.73	3.74	3.88	4.59	4.26	4.51
Cheese		4.88	4.11	3.89	3.72	4.50	3.58	3.75
Meat and meat products		29.65	32.87	34.03	32.40	37.93	34.37	35.81
Fish		5.08	4.74	4.46	4.13	8.53	5.33	6.61
Eggs	(no)	1.51	1.75	2.06	2.38	2.57	2.53	2.91
Fats and oils		7.57	8.11	8.19	8.94	11.33	8.81	11.72
Sugar and preserves		5.02	5.66	6.29	7.72	11.79	8.52	12.88
Fruit		46.19	35.24	27.06	24.31	54.19	26.84	37.46
Vegetables		66.17	72.39	77.73	79.83	89.57	85.80	86.33
Cereals (incl. bread)		48.13	49.49	51.36	53.20	54.91	53.95	57.11
Beverages		2.21	2.04	2.17	2.04	3.55	2.61	3.82
Other foods		25.49	14.08	11.70	11.64	16.85	11.02	10.34
Soft drinks	(fl oz)	28.96	28.47	26.66	23.14	20.95	21.52	14.69
Alcoholic drinks	(cl)	42.05	38.54	28.97	21.64	41.23	20.18	14.10
Confectionery	(g)	52.13	54.52	50.61	43.82	55.50	46.55	48.39

---

[3] Respondents are often reluctant or unable to give precise details of the income of the head of household or, where appropriate, the principal earner. In such cases they are asked to state the income group in which it lies. Details of the sample of households in each income group for 1992 are given in Appendix A.

Table 2.20 *continued*

per person per week

	INCOME GROUP (a)						
	Gross weekly income of head of household						
	Households with one or more earners				Households without an earner		OAP
	A	B	C	D	E1	E2	
EXPENDITURE				(pence)			
Milk and cream	166.64	144.68	136.73	128.41	172.33	133.14	157.25
Cheese	62.89	48.02	42.69	40.40	56.27	40.04	42.43
Meat and meat products	367.73	356.15	331.32	290.36	402.08	304.11	351.54
Fish	87.93	66.30	57.01	49.97	128.78	65.21	92.39
Eggs	15.64	16.10	17.92	19.75	24.56	22.63	26.47
Fats and oils	36.93	36.62	34.27	33.83	53.44	35.17	51.41
Sugar and preserves	16.20	15.02	15.59	17.91	32.39	21.00	32.63
Fruit	162.72	112.30	84.19	74.08	177.63	81.96	114.09
Vegetables	222.89	195.97	178.46	167.05	201.40	165.60	148.86
Cereals (incl. bread)	260.42	235.67	218.19	199.44	247.65	208.38	222.47
Beverages	44.73	39.31	40.64	35.87	62.83	43.80	63.14
Other foods	94.26	68.24	58.45	53.22	82.68	54.32	52.19
Total food	**£15.39**	**£13.34**	**£12.15**	**£11.10**	**£16.42**	**£11.75**	**£13.55**
Soft drinks	57.46	49.21	44.04	37.85	35.98	34.54	23.58
Alcoholic drinks	159.41	103.39	73.93	50.20	191.49	68.91	58.08
Confectionery	26.11	25.78	23.16	19.69	27.06	20.16	21.40
Total food and drink	**£17.82**	**£15.13**	**£13.57**	**£12.18**	**£18.97**	**£12.99**	**£14.58**

(a) definition: A £520 and over, B £280 and under £520, C £140 and under £280, D less than £140, E1 £140 or more, E2 less than £140

(b) except where otherwise stated

income, and sugar, preserves and beverages. However the quantities of most of the foods consumed did not tend to rise with income, and consumption of eggs, fats and oils, sugar and preserves, vegetables and cereals were all lower in households with higher head of household income.

Figures 2.21 to 2.23 illustrate, for a selection of foods, variations of expenditure and consumption between income groups. It is apparent that people in the E1 and OAP groups, who tend to be older, consumed most carcase meat, particularly bacon and ham. In households with an earner, consumption of meat appeared to decline with income, except for the lowest earning group, but expenditure clearly rose with income, especially for poultry, implying that the higher earners bought the more expensive cuts. For fish, consumption and expenditure patterns are closely related. Expenditure on confectionery also increased with income, although consumption of the different products was more variable; consumption of chocolate confectionery rose with income but consumption of mints and boiled sweets declined.

Figure 2.21
Consumption and expenditure for carcase meat by income group, 1992

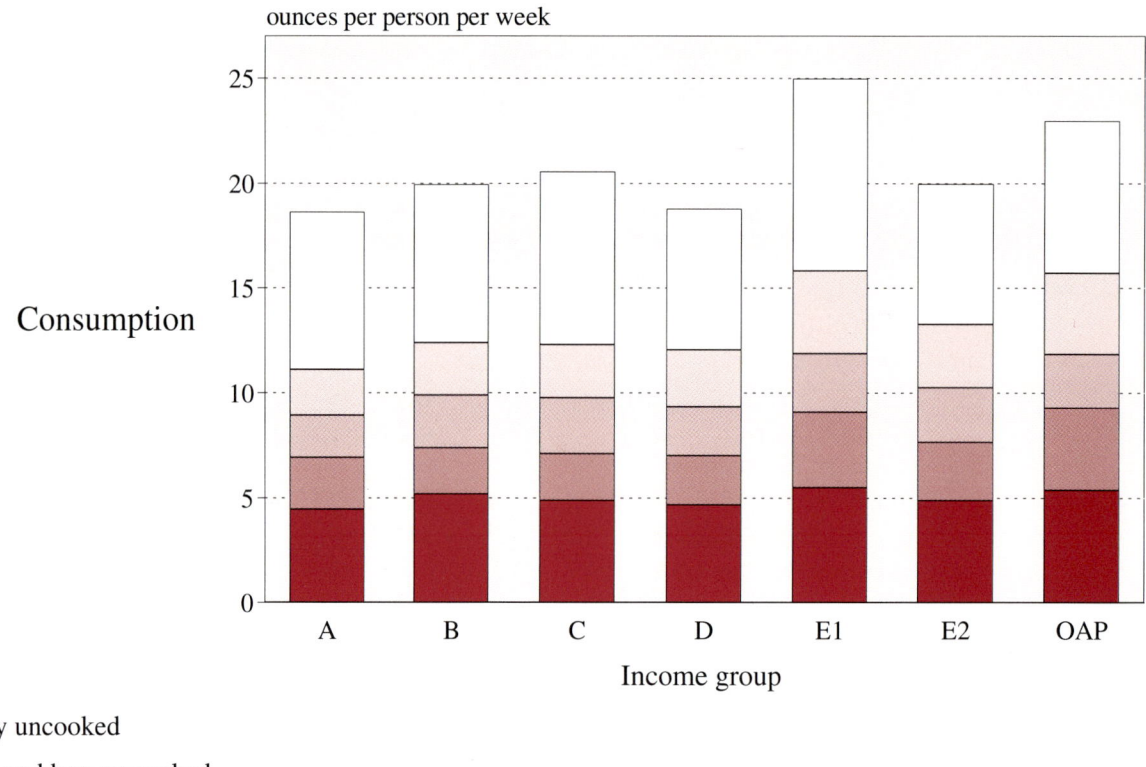
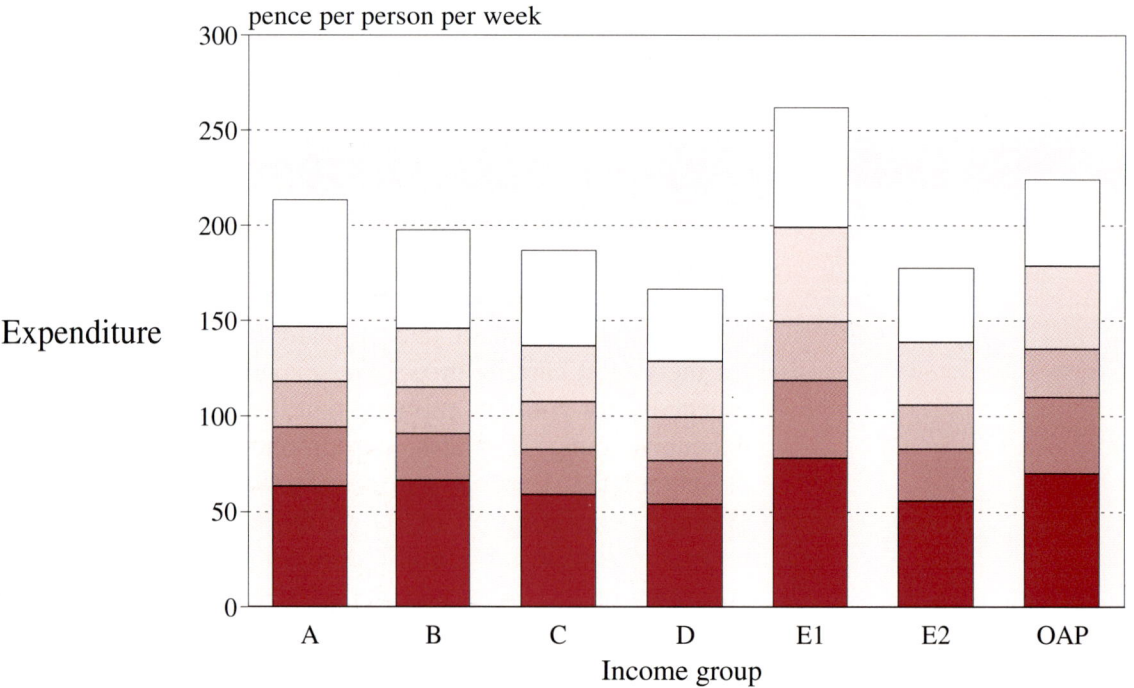

Figure 2.22
Consumption and expenditure for fish by income group, 1992

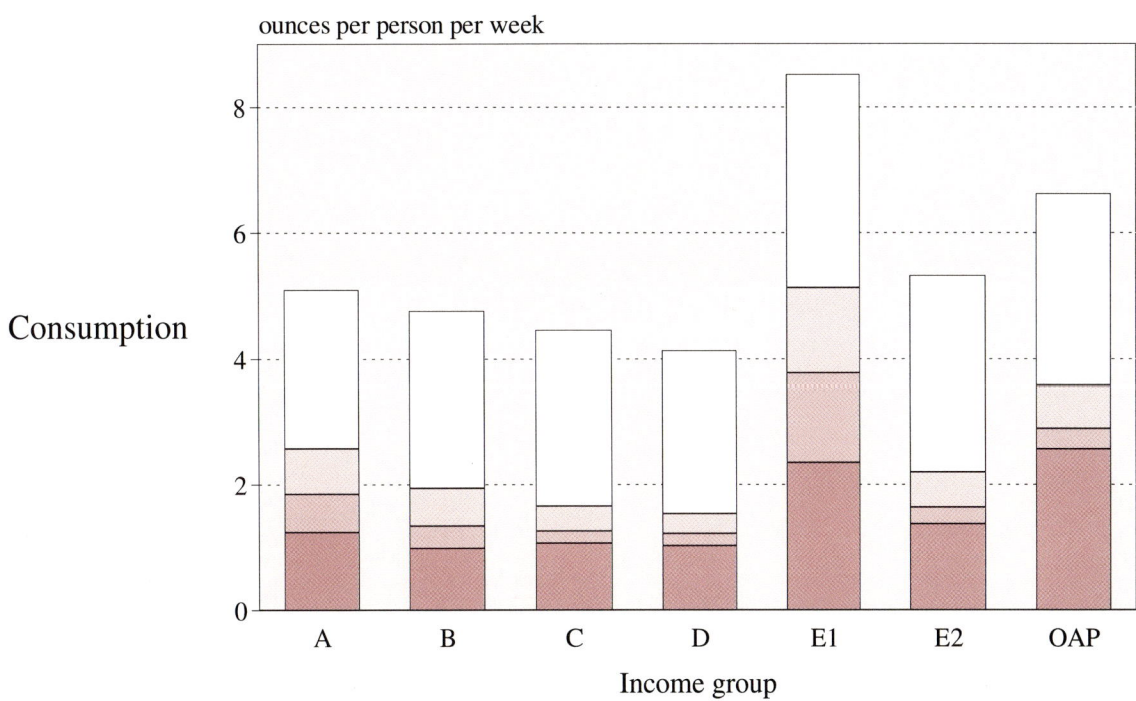

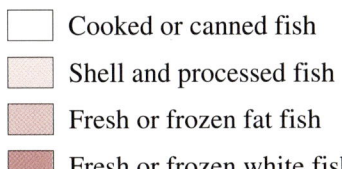
Cooked or canned fish
Shell and processed fish
Fresh or frozen fat fish
Fresh or frozen white fish

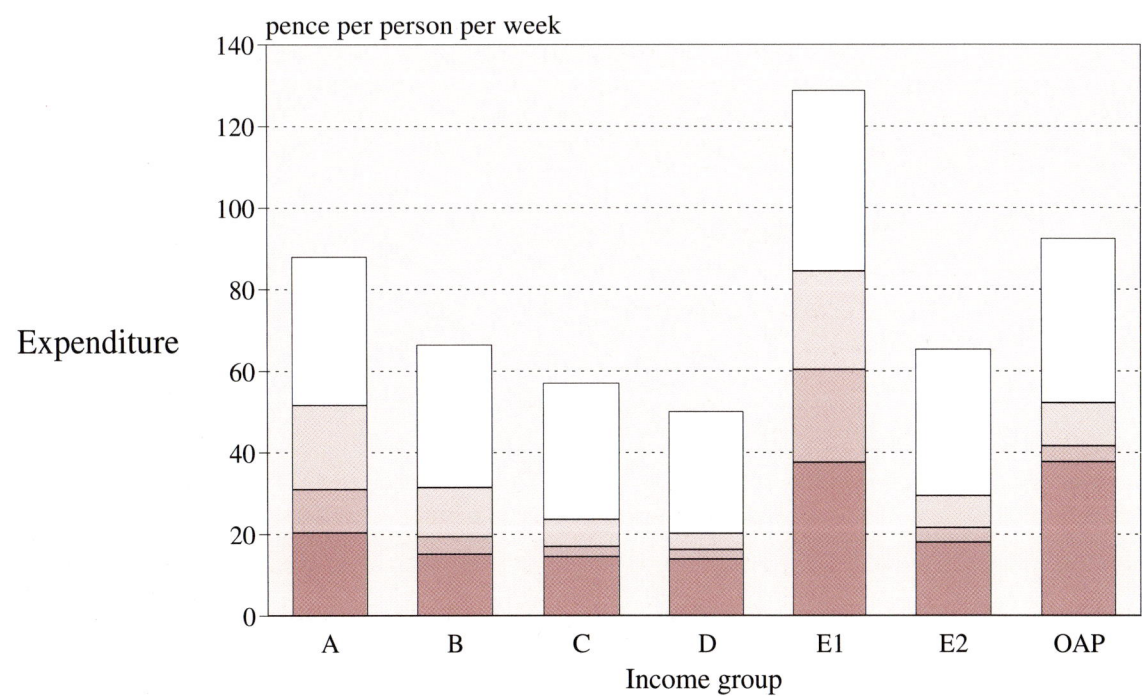

Figure 2.23
Consumption and expenditure for confectionery brought home, by income group, 1992

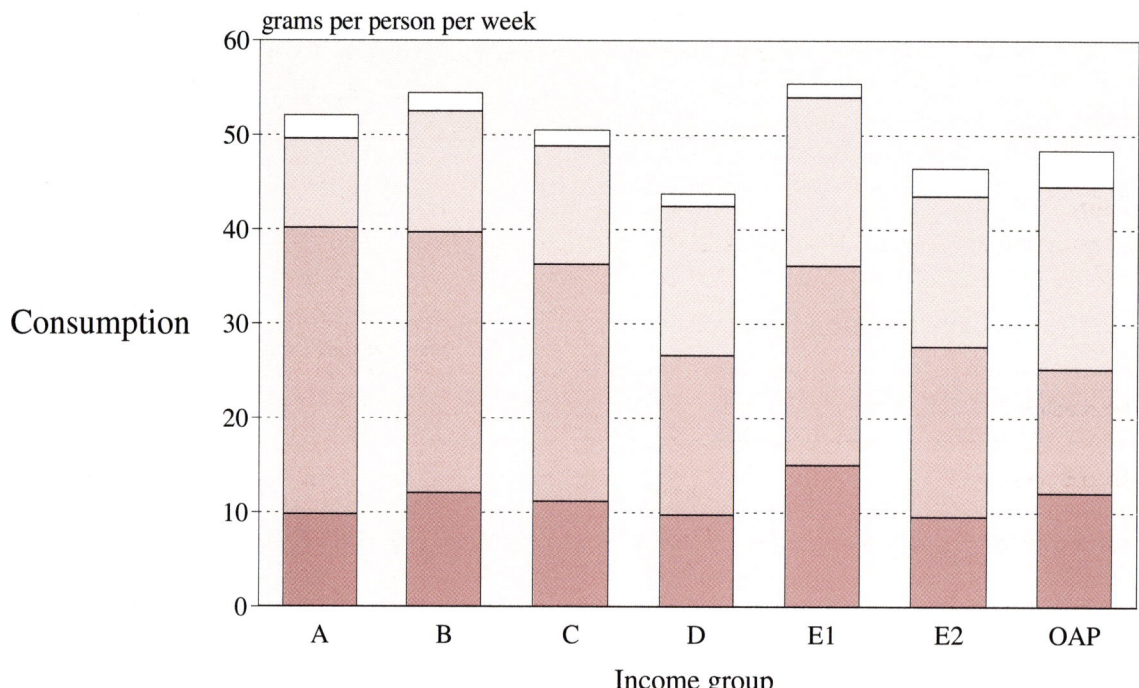

- Other
- Mints and sweets
- Chocolate coated and filled bars
- Solid chocolate

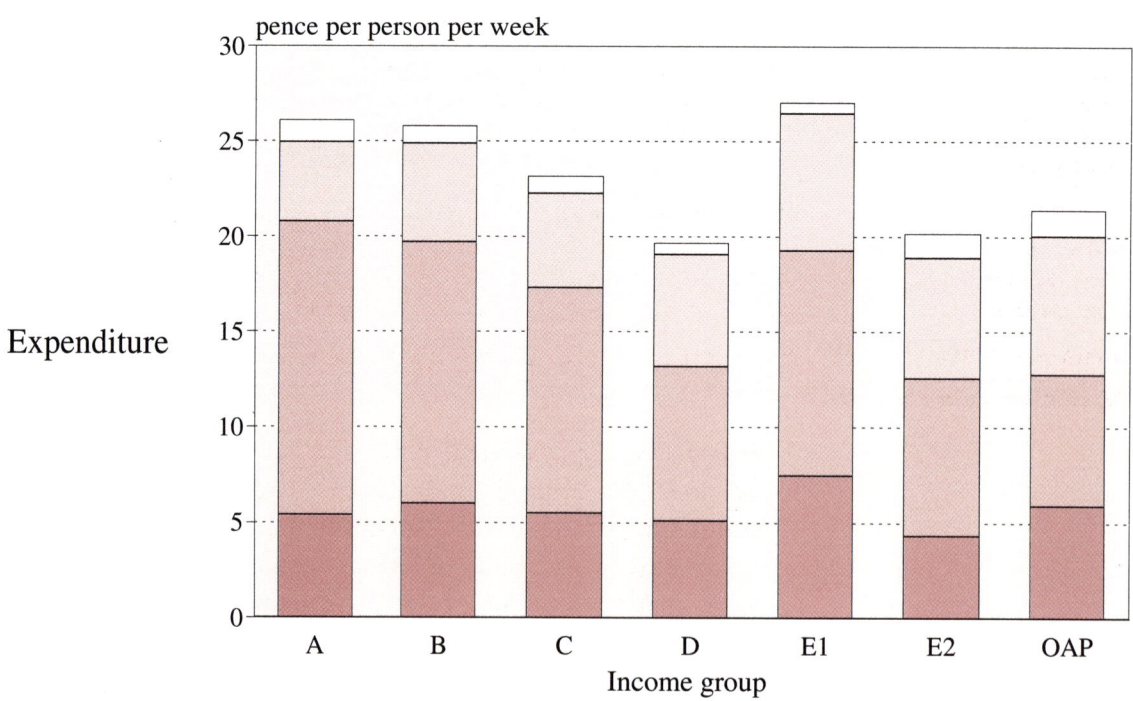

Analysis by Household Composition

This analysis shows the extent to which consumption and expenditure for selected foods is affected by the size and composition of the household. Table 2.24 provides a broad summary; more detailed information on expenditure for individual foods is given in Appendix Table B7.

Table 2.24
Consumption of selected foods by household composition, 1992

ounces per person per week[a]

		Households with									
No of adults		1		2					3	3 or more	4 or more
No of children		0	1 or more	0	1	2	3	4 or more	0	1 or 2	3 or more	0
Milk and cream (pt or eq pt)		4.55	3.78	4.13	3.95	3.69	3.76	3.69	3.90	3.54	3.68	3.40
Cheese		4.74	2.89	4.63	3.85	3.66	2.82	3.06	4.64	3.98	3.05	3.47
Carcase meat		10.15	7.41	12.64	8.82	7.92	6.67	7.08	11.94	10.73	6.55	11.63
Other meat and meat products		24.99	21.21	27.27	23.09	19.93	19.05	21.22	27.52	22.18	20.64	21.25
Fish		6.76	3.63	6.96	3.99	3.79	3.31	3.06	5.70	3.93	2.94	3.76
Eggs (no)		2.83	1.95	2.49	1.88	1.67	1.55	1.56	2.36	1.69	1.54	2.00
Fats		10.10	6.76	10.24	7.09	7.18	7.24	7.99	9.55	8.56	7.89	9.90
Sugar and preserves		9.71	6.10	9.32	5.84	4.47	4.94	5.51	8.12	6.71	7.52	7.13
Potatoes		32.36	30.08	35.99	29.87	27.02	23.39	30.81	36.78	32.03	32.55	36.80
Fresh green vegetables		11.68	4.60	12.32	7.28	6.15	5.26	5.13	11.30	7.71	5.57	7.86
Other fresh vegetables		21.12	10.98	21.82	15.49	12.70	11.09	12.70	19.46	14.65	13.67	15.17
Processed vegetables		18.21	21.65	20.34	21.63	19.66	18.15	22.57	20.72	19.58	17.80	19.83
Fresh fruit		30.12	13.72	28.70	17.29	18.35	16.19	13.35	23.70	18.51	13.24	16.93
Fruit juices		9.35	5.56	8.45	8.54	7.57	5.92	8.59	7.65	7.60	4.47	7.07
Other fruit and fruit products		4.59	1.43	4.77	2.19	2.26	1.80	1.58	3.65	2.42	2.89	2.32
Bread		30.48	23.25	29.75	25.17	23.08	23.12	26.87	29.16	25.96	21.95	25.95
Other cereals		27.71	23.91	26.49	23.52	23.66	24.21	24.44	24.46	22.24	25.69	27.26
Tea		1.84	1.00	1.86	1.08	0.82	0.89	1.26	1.64	1.32	0.90	1.50
Coffee		0.85	0.51	0.84	0.56	0.44	0.46	0.40	0.69	0.58	0.44	0.57
Other beverages		0.48	0.21	0.45	0.34	0.19	0.31	0.42	0.50	0.22	0.23	0.39
Other foods		13.73	8.83	15.64	12.85	12.16	10.60	10.43	19.96	12.04	8.73	12.21
Total food		**£15.26**	**£9.68**	**£15.61**	**£12.46**	**£11.01**	**£9.76**	**£9.74**	**£14.50**	**£11.73**	**£9.67**	**£12.22**
Soft drinks (fl oz)		21.14	27.99	21.97	28.87	29.47	26.81	28.73	22.38	27.70	30.35	21.41
Alcoholic drinks (cl)		38.56	10.45	44.40	33.21	25.46	16.57	22.68	31.58	20.62	7.42	24.30
Confectionery (g)		50.11	42.63	54.25	55.48	52.73	63.38	41.88	45.58	45.34	29.23	32.63
Total food and drink		**£17.16**	**£10.64**	**£17.75**	**£14.03**	**£12.37**	**£10.88**	**£10.73**	**£16.06**	**£12.91**	**£10.47**	**£13.49**

(a) except where otherwise stated

It is to be expected that per capita consumption and expenditure will be lower in larger households and those with children as a result of economies of scale and the lower needs of children. They will also be affected by associated differences in incomes (both total and per capita), meal patterns and degree of eating-out.

Comparing firstly households without children, Figure 2.25 shows that expenditure per person was highest in households with one or two adults and then declined with increasing household size. Per capita expenditure on meat, fish and eggs and vegetables was highest in households with two adults only and declined markedly, together with spending on milk products, bread and fruit, in the larger households.

Figure 2.25
Expenditure on main food groups per person by number of people in adult-only households, 1992

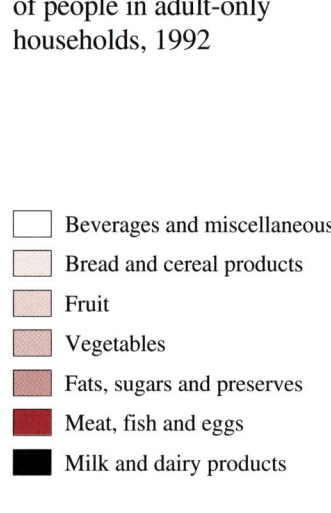
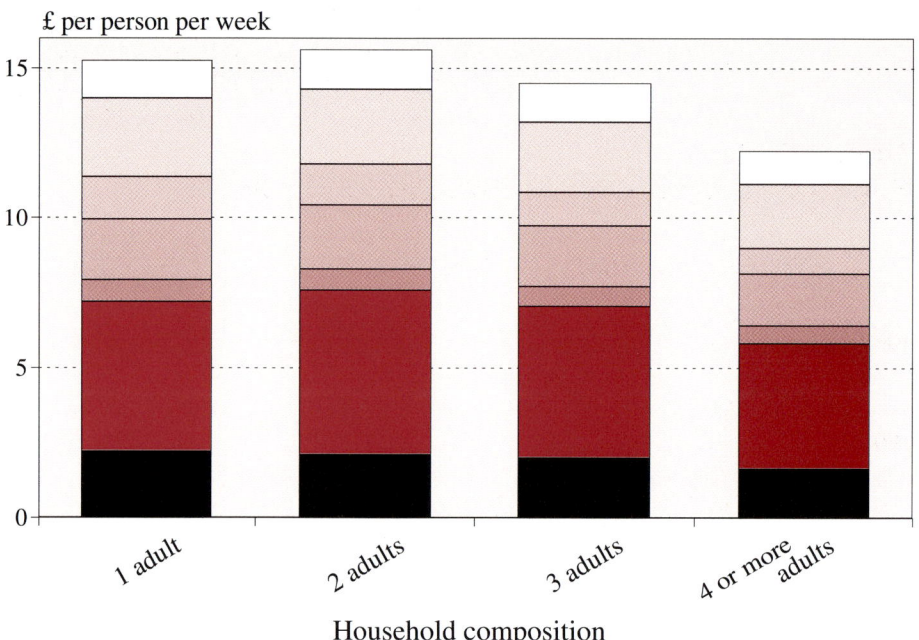

Turning to households with children, the reduction in total expenditure per person in two-adult households is most pronounced as the first child is added; the addition of further children does not affect expenditure to quite the same degree (Figure 2.26). Per capita expenditure on fish, fats, sugar and preserves, fruit and beverages is reduced by more than a third with the addition of a child to a two adult household.

Figure 2.26
Expenditure on main food groups per person by number of children in 2-adult households, 1992

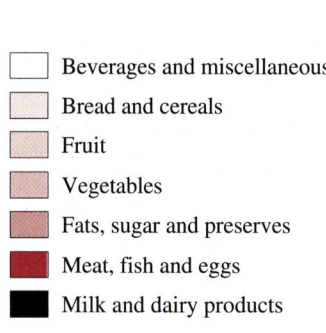
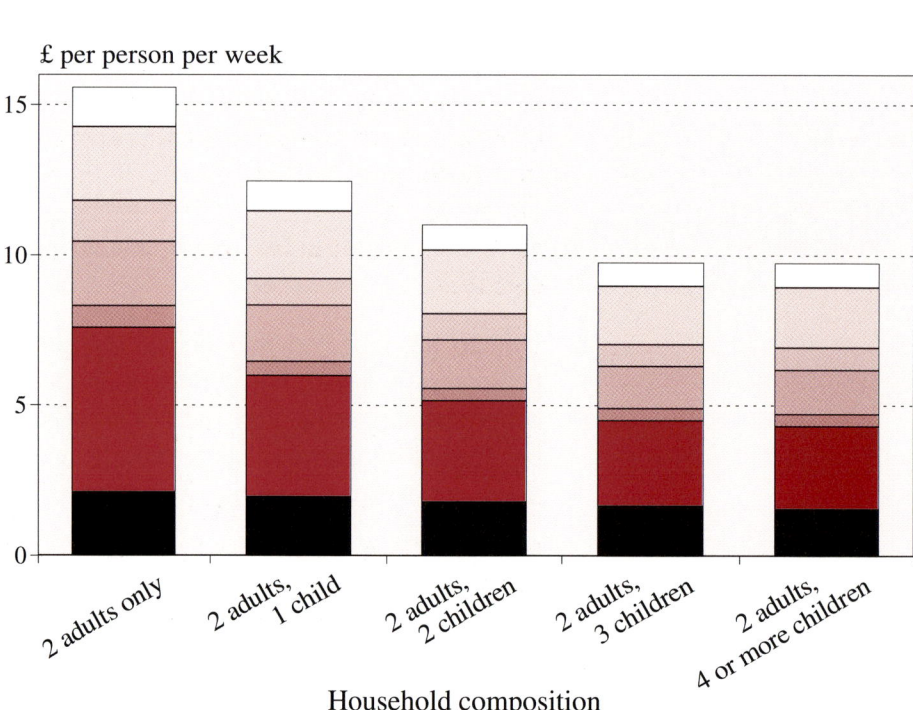

The pattern of reductions in per capita expenditure accompanying the addition of extra household members is apparent for fresh green and other fresh vegetables (Figure 2.27). However the overall pattern is less clear for processed vegetables where there is some tendency for per capita expenditure to rise with the number of people in adult only households, and with the addition of a single child, but then fall as more children are present in the household.

Figure 2.27
Expenditure on vegetables by household composition, 1992

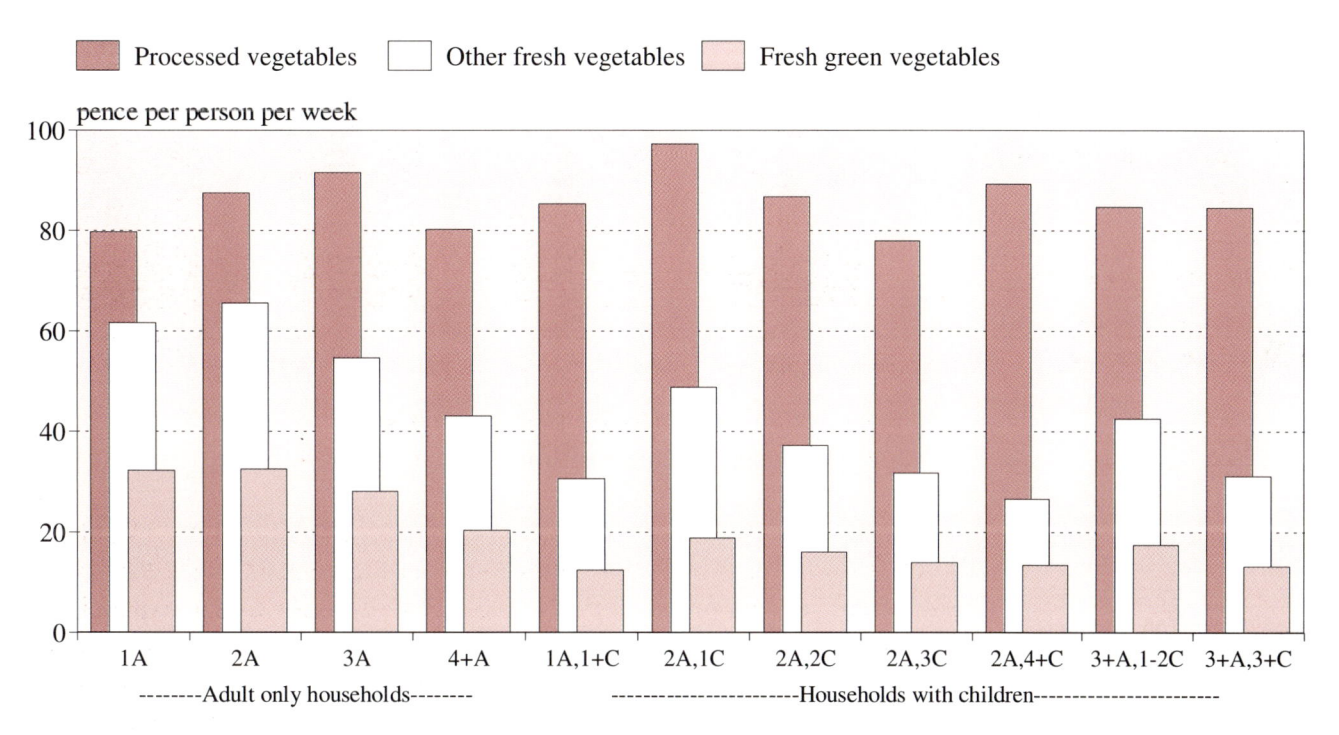

Expenditure per person on concentrated and unconcentrated soft drinks is only marginally affected by numbers of adults (Figure 2.28). However expenditure on concentrated soft drinks rises sharply with an increase in the number of children whilst an opposite effect is seen for unconcentrated soft drinks. Expenditure on fruit juices tends to decline with both larger numbers of adults and larger numbers of children.

The level of consumption of confectionery in the home, like that of soft drinks, is much affected by the presence of children in the household (Figure 2.29). Consumption of chocolate-coated or filled sweets is greatest in the 2-adult households with children, although this is not the case for solid chocolate. Mints, fudge and boiled sweets tend to be favoured by the adult-only households, with the highest consumption being in those with just 1 or 2 adults; though the limitation in coverage to items 'brought home' may be particularly important here.

Figure 2.28
Expenditure on soft drinks brought home and fruit juice by household composition, 1992

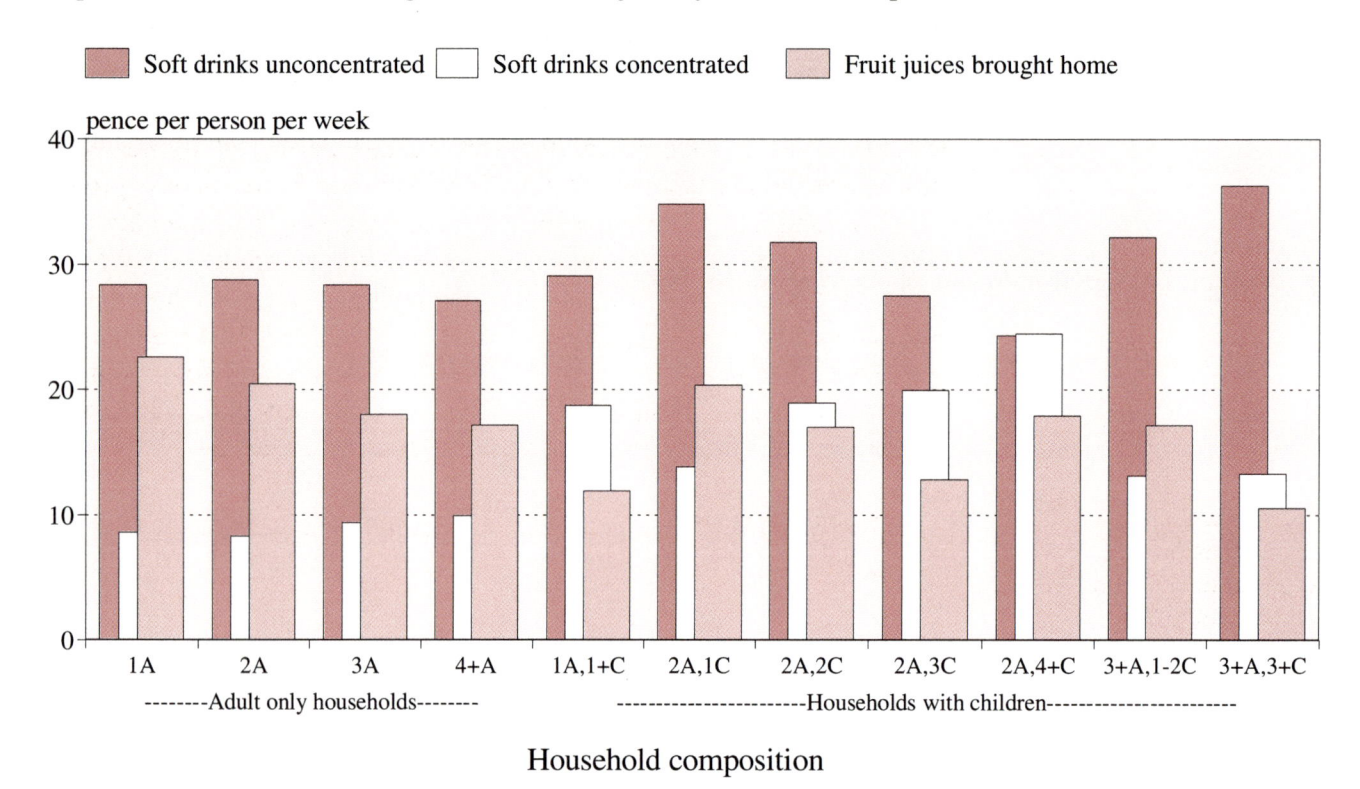

Figure 2.29
Consumption of confectionery brought home, by household composition, 1992

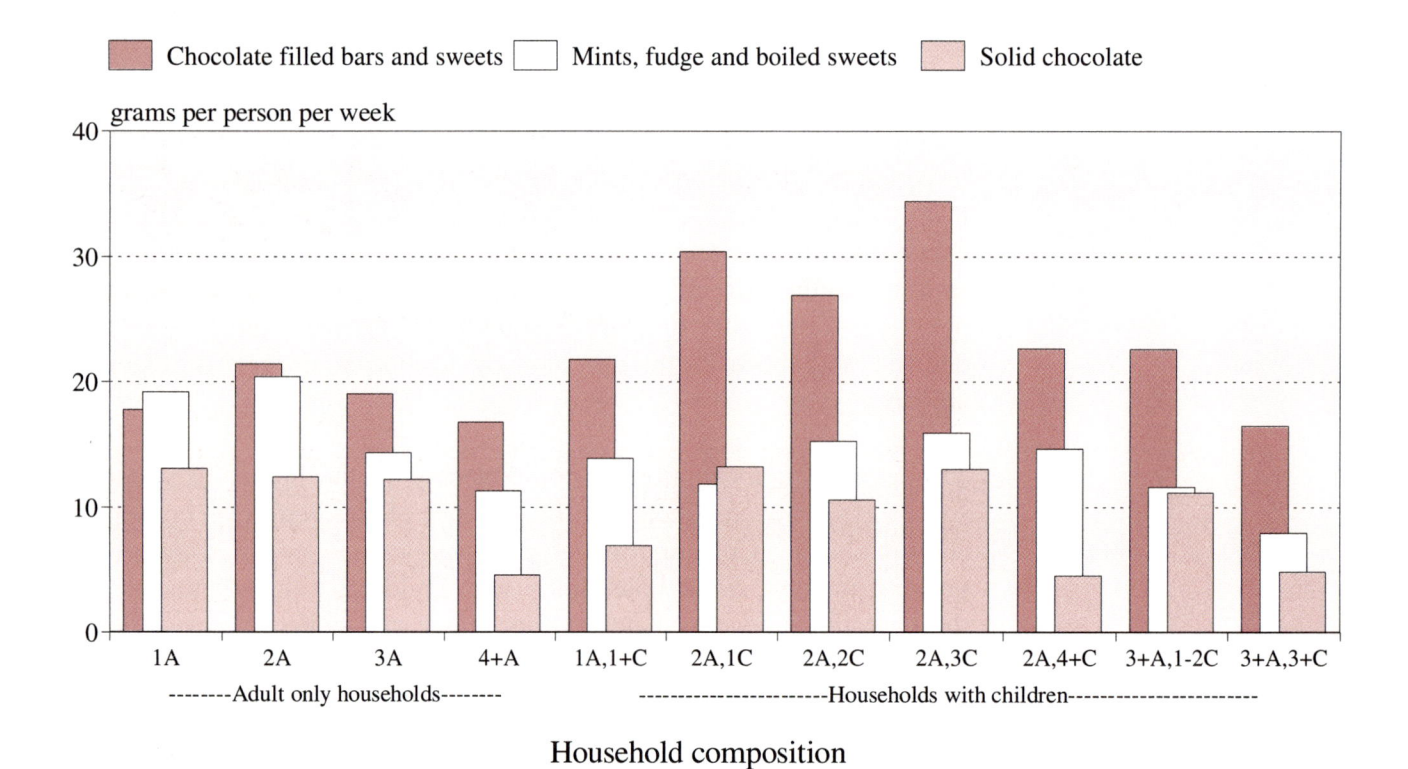

Analysis by Household Composition and Income

Average expenditure on household food per head showed greater variation between households of different composition than between households in different income groups with the same composition (Figure 2.30). Income-related differences appeared to be more pronounced in households with children, particularly for income groups B, C, D and E2. The average per person increase in expenditure in adult only households between those in income groups D and E2 and those in group B was £1.57. A comparable increase in households with children was £2.93. The lowest average weekly food expenditure was £6.83 per person for families with two adults and four or more children in income group D or E2. The highest expenditure was £18.75 per person in adult-only households in income group A.

Figure 2.30
Total household food expenditure per head by certain household composition groups within income groups, 1992

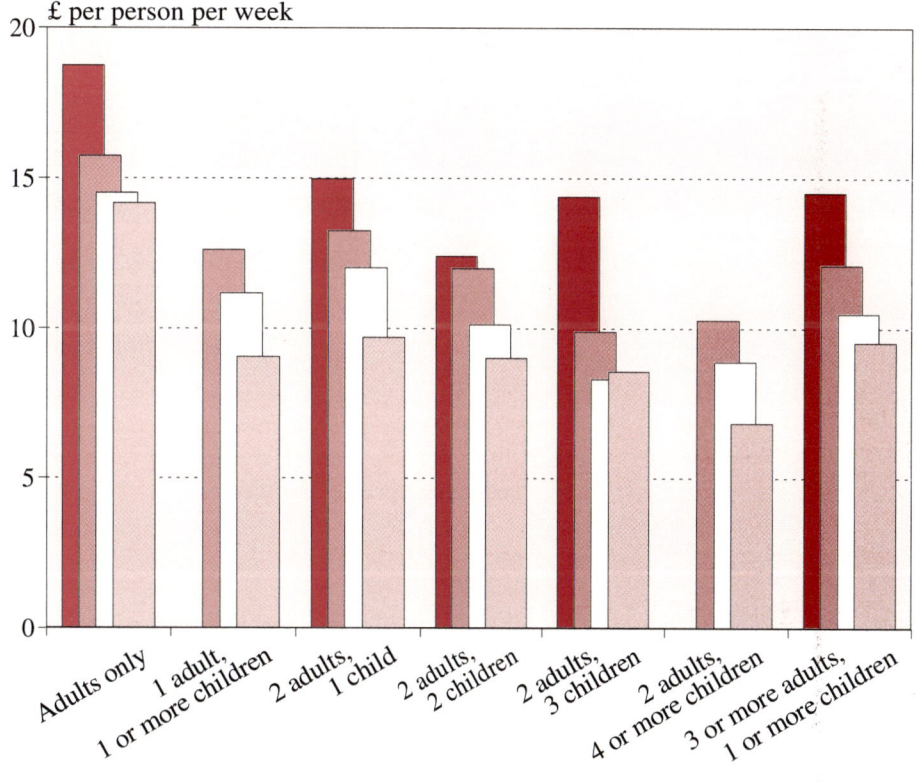

Income group
- A
- B
- C
- D+E2

The variation in expenditure by household composition and income group is seen in more detail in a selection of foods (Figure 2.31). Income had little effect on expenditure on eggs in households with 2 adults and any number of children. However expenditure on eggs tended to rise with decreasing income in households without children. There was a marked income effect on total expenditure on fruit in households of all compositions. Total expenditure on milk varies little between households of different income groups within the same composition. However far more is spent on low fat milks by households in higher income groups and correspondingly less is spent on wholemilk; this pattern is consistent for almost all household compositions.

Figure 2.31
Expenditure on selected foods by certain household composition groups within income groups, 1992

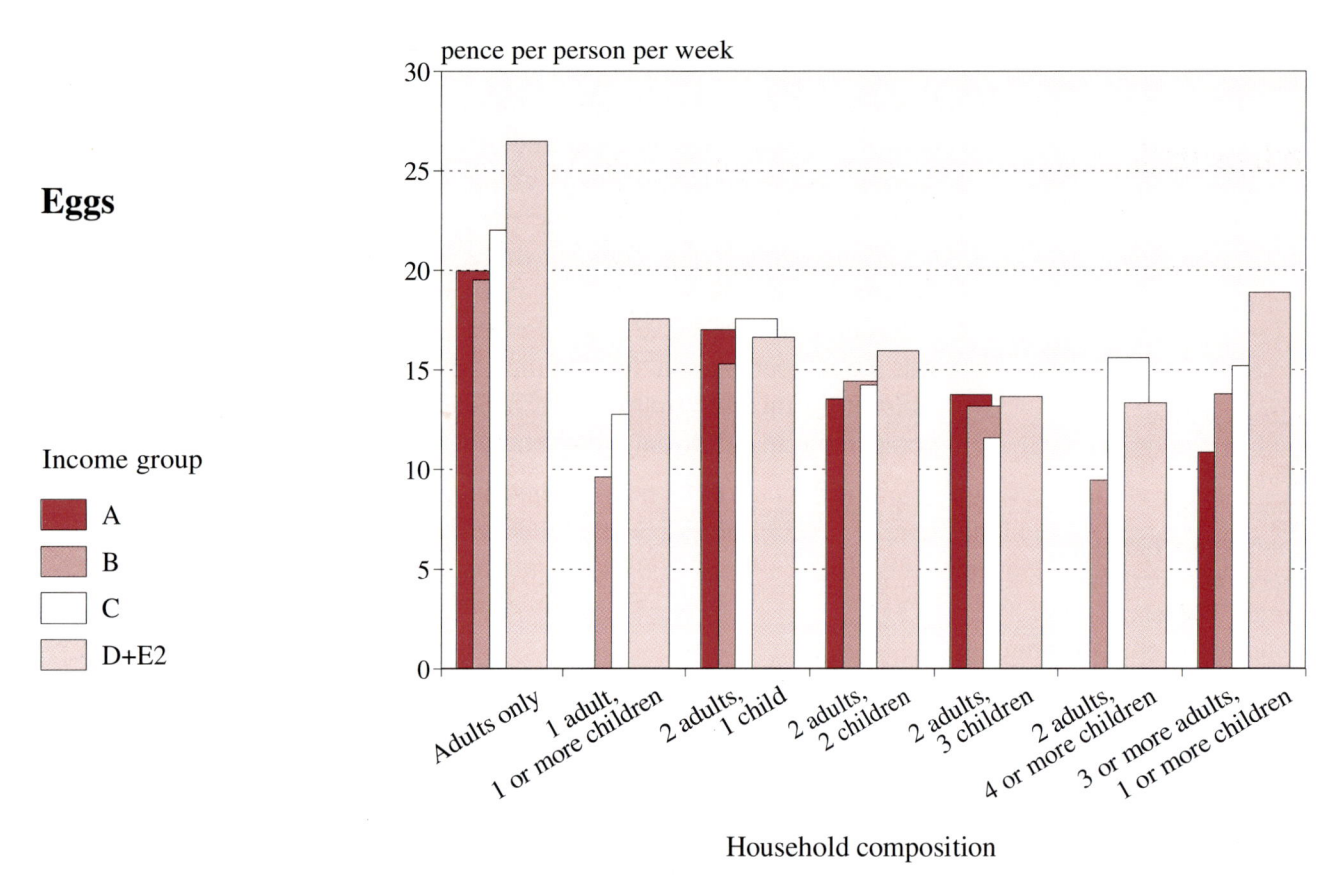

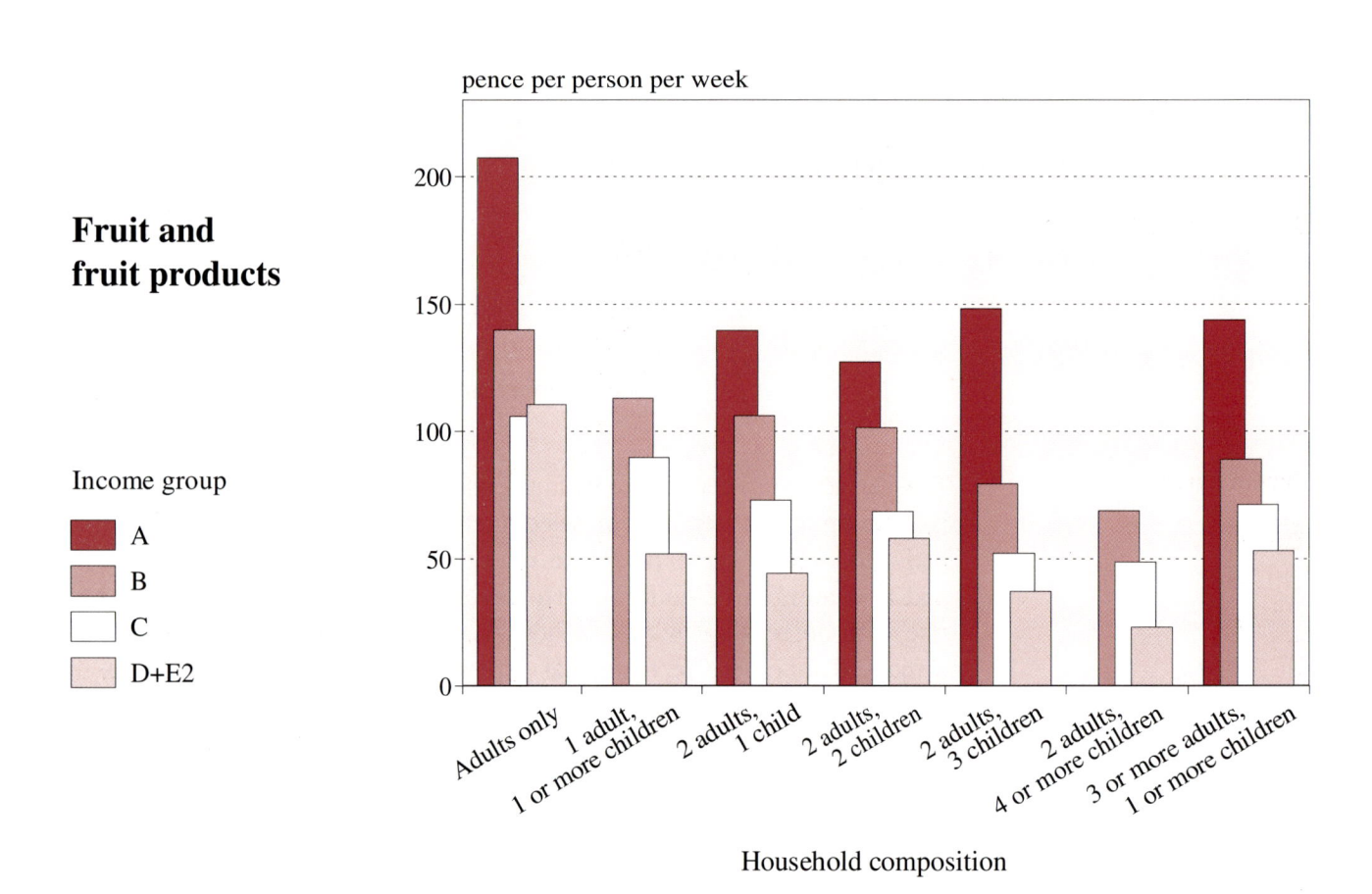

Eggs

Income group
- A
- B
- C
- D+E2

Fruit and fruit products

Income group
- A
- B
- C
- D+E2

Figure 2.31 continued

Whole milk

Income group
- A
- B
- C
- D+E2

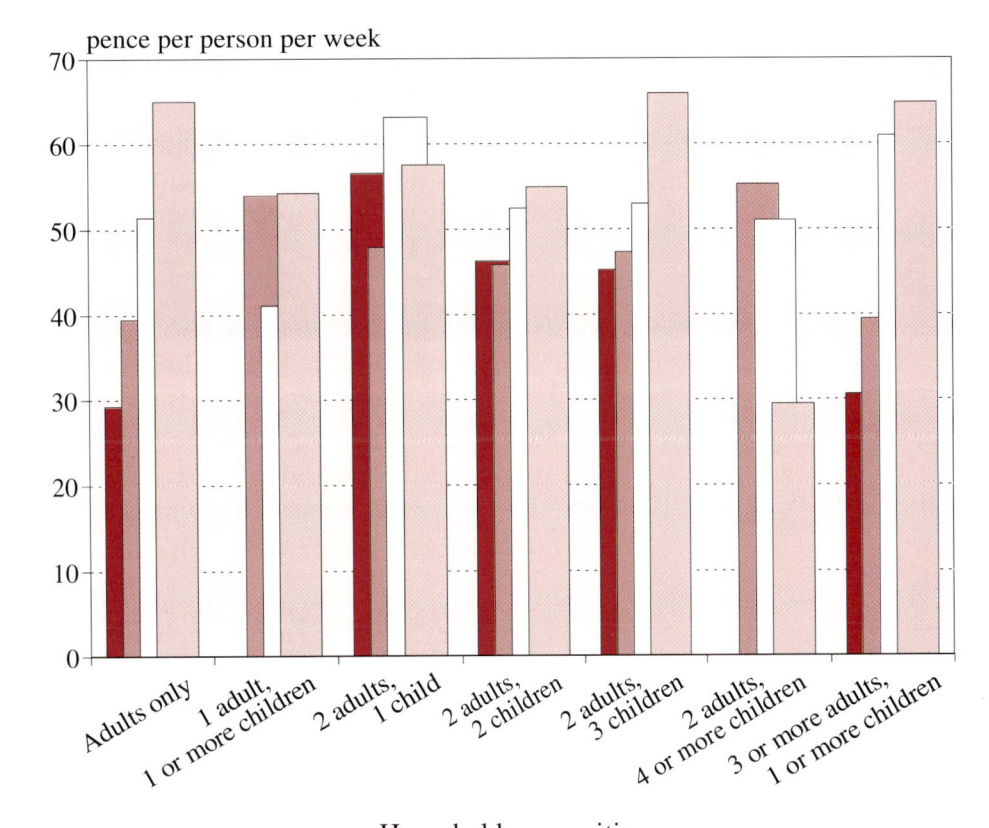

Low fat milk

Income group
- A
- B
- C
- D+E2

29

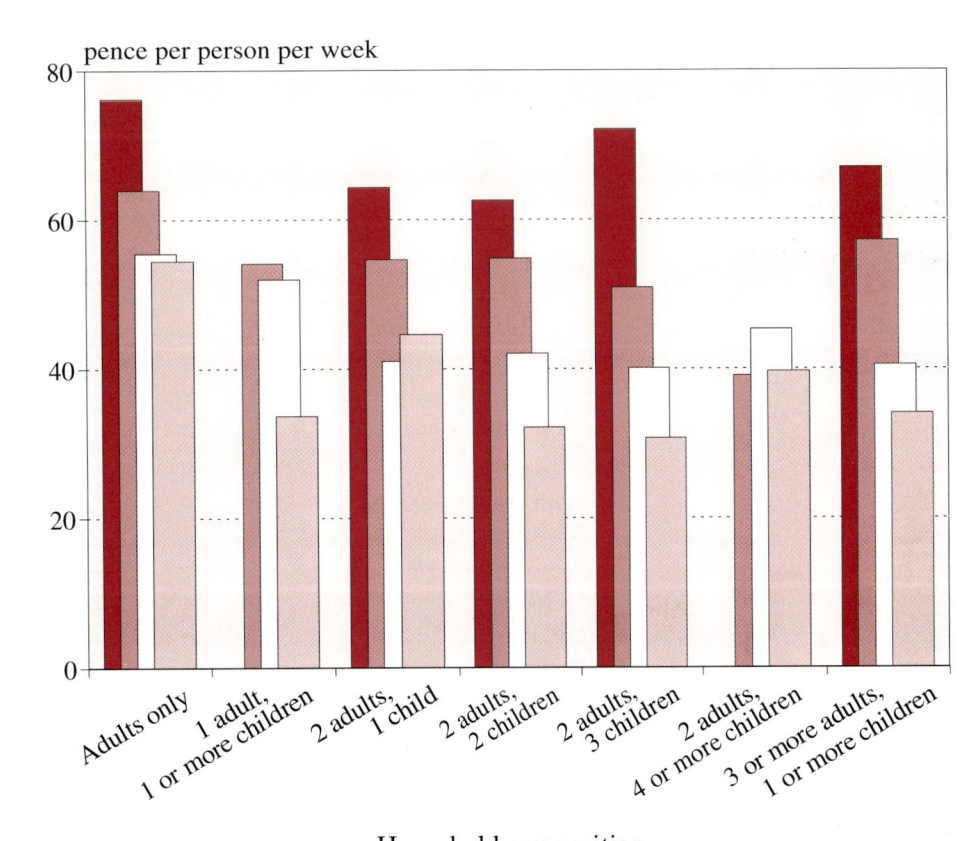

Analysis by Age of Main Diary Keeper

The main diary-keeper is the member of the household who is most responsible for the purchase of household food and the provision of meals. Factors such as the composition of the household and the level of income affect decisions regarding what items are to be bought but these are also often much influenced by other factors such as the age of the diary-keeper.

Expenditure per person tended to increase with age of main diary-keeper up to retirement age, being almost half as much again for those in the 55 - 64 age group than for those in the under 25 age group (Table 2.32). This was also the pattern for expenditure on most types of food with the exception of expenditure on processed vegetables, soft drinks and confectionery, for which the presence of children has been shown to have a significant effect.

The main diary-keepers under 25 purchased more processed vegetables for their households, whereas the consumption of fresh green and other

Table 2.32
Consumption and expenditure for selected foods by age of main diary-keeper, 1992

per person per week

		\multicolumn{7}{c}{Age of main diary-keeper}
		under 25	25-34	35-44	45-54	55-64	65-74	75 plus
CONSUMPTION		\multicolumn{7}{c}{(ounces)[a]}
Total milk and cream (pt or eq pt)		3.4	3.6	3.7	4.0	4.5	4.4	4.6
of which:								
Wholemilk		1.5	1.7	1.5	1.6	1.9	2.1	2.6
Low fat milk		1.2	1.4	1.8	2.0	2.0	1.8	1.4
Cheese		3.8	3.4	4.0	4.6	4.7	4.3	3.0
Carcase meat		7.1	7.0	9.4	12.2	13.8	12.8	11.0
Other meat and meat products		20.2	20.6	21.9	27.0	29.4	25.4	22.1
Fish		3.8	3.8	4.1	5.7	7.2	7.0	6.7
Eggs	(no)	1.7	1.6	1.8	2.3	2.8	2.9	2.7
Fats		6.6	6.4	8.0	10.0	11.2	11.6	10.5
of which:								
Butter		0.7	0.9	1.2	1.7	2.3	2.3	2.6
Margarine		1.8	2.0	2.8	3.1	3.2	3.9	3.8
Low fat spreads		1.8	1.5	1.6	2.1	2.2	2.1	1.8
Sugar and preserves		4.7	4.3	5.7	8.2	10.2	12.2	12.4
Potatoes		23.5	24.1	30.6	35.9	43.1	36.0	40.5
Fresh green vegetables		4.2	5.4	7.3	11.3	13.7	13.5	12.0
Other fresh vegetables		11.6	12.4	16.0	19.7	23.9	20.4	15.9
Processed vegetables		24.6	20.2	20.9	21.1	19.9	15.5	12.5
Fruit and other fruit products		21.0	24.4	31.3	39.0	42.2	41.2	38.9
Cereals		43.4	44.4	50.5	56.2	61.8	58.4	52.0
Beverages		1.3	1.5	2.0	2.8	3.7	3.5	3.4
Soft drinks	(fl oz)	27.4	25.0	29.7	28.2	23.1	15.3	13.4
Alcoholic drinks	(cl)	36.8	35.9	28.3	30.0	35.0	24.5	14.4
Confectionery	(g)	39.8	49.5	53.1	51.0	53.9	56.8	43.4

(a) except where otherwise stated

Table 2.32 *continued*

per person per week

	\multicolumn{7}{c}{Age of main diary-keeper}
	under 25	25-34	35-44	45-54	55-64	65-74	75 plus
EXPENDITURE				(pence)			
Total milk and cream	115.5	131.8	139.8	150.3	165.4	158.5	160.4
of which:							
Wholemilk	46.5	51.8	45.8	49.2	59.8	68.1	88.2
Low fat milk	36.5	40.9	55.8	62.1	63.4	56.8	45.8
Cheese	41.2	38.8	45.8	52.9	54.9	50.2	35.8
Carcase meat	74.9	74.6	103.3	143.7	163.6	152.9	127.9
Other meat and meat products	201.8	202.3	210.6	259.6	277.1	236.8	205.8
Fish	47.2	49.8	54.4	79.0	100.9	101.0	95.2
Eggs	15.7	13.9	15.9	20.9	25.9	27.0	24.8
Fats	26.4	27.0	33.3	43.6	50.4	51.9	48.8
of which:							
Butter	4.7	5.9	8.0	11.7	16.3	16.3	17.9
Margarine	6.0	7.2	9.7	11.4	12.2	14.4	14.4
Reduced and low fat spreads	9.4	7.8	8.1	10.4	11.2	10.8	9.0
Sugar and preserves	11.3	10.7	14.1	21.3	27.2	31.5	34.4
Potatoes	19.8	20.5	23.0	29.7	32.1	29.0	28.7
Fresh green vegetables	13.1	16.2	19.1	29.8	34.3	31.1	25.5
Other fresh vegetables	40.8	40.3	47.6	58.9	65.5	49.2	36.7
Processed vegetables	103.6	92.0	92.4	92.3	80.7	58.8	46.0
Fruit and other fruit products	64.1	78.2	99.5	125.7	135.0	128.5	121.4
Cereals	206.0	203.5	223.1	245.0	257.5	236.9	214.0
Beverages	25.3	28.9	36.7	53.6	65.3	52.3	58.7
Miscellaneous (expenditure only)	62.7	57.5	61.0	74.3	76.5	60.9	53.4
Total of above	**£10.69**	**£10.86**	**£12.20**	**£14.81**	**£16.12**	**£14.62**	**£13.18**
Soft drinks	46.8	43.7	50.4	47.6	38.6	25.0	22.9
Alcoholic drinks	84.5	83.3	77.6	100.2	125.3	105.1	72.3
Confectionery	18.9	23.4	24.7	24.1	23.3	26.0	19.6
Total expenditure	**£12.19**	**£12.36**	**£13.72**	**£16.52**	**£17.99**	**£16.19**	**£14.32**

fresh vegetables was much higher where the diary-keeper was over 55. The older age groups tended to buy more butter and margarine; consumption of reduced and low fat spreads in households where the diary-keeper was under 25 was twice as high as butter. Although per capita spending on alcoholic drinks brought home peaked, at £1.25 per person, in the 55 to 64 age group, consumption by volume was higher where the diary-keeper was under 34.

Analysis by Ownership of Microwaves and Freezers

Some 87 per cent of the households participating in the Survey in 1992 owned a freezer, while 61 per cent owned microwaves. Expenditure by households who owned one or both appliances was marginally higher than for those who owned neither (Table 2.33).

A household's purchasing pattern for specific items may be affected by ownership of a freezer or a microwave though patterns between broad groups of food expenditure are more likely to be related to other underlying factors such as income group distribution.

Table 2.33
Expenditure on selected foods by ownership of microwave or freezer, 1992

per person per week

	Microwave only	Freezer only	Both	Neither
EXPENDITURE		(pence)		
Total milk and cream	131.0	147.5	143.0	137.5
Cheese	44.8	46.9	45.9	44.6
Carcase meat	102.1	116.9	113.1	111.2
Other meat and meat products	242.3	221.3	230.2	200.1
Fish	74.1	71.5	66.4	72.0
Eggs	24.4	19.8	17.7	23.6
Fats	36.5	39.3	36.3	41.2
Sugar and preserves	19.2	21.1	16.1	28.0
Potatoes	25.9	25.5	24.6	28.5
Fresh green vegetables	24.1	23.7	22.2	27.4
Other fresh vegetables	46.0	48.8	49.2	49.3
Frozen vegetables	17.5	23.3	25.0	12.6
Other vegetables	81.3	59.5	64.6	55.2
Fresh fruit	61.7	72.4	68.1	71.4
Other fruit products	29.2	36.6	34.7	34.2
Bread	81.4	70.7	67.3	78.0
Other cereals	171.5	151.3	158.4	150.4
Beverages	44.7	43.7	42.4	47.4
Miscellaneous	61.5	62.4	65.9	52.9
Total food	**£13.19**	**£13.02**	**£12.91**	**£12.67**
Soft drinks	39.4	39.4	46.3	30.4
Alcoholic drinks	87.7	96.6	90.7	73.0
Confectionery	26.0	23.0	23.7	23.9
Total expenditure	**£14.72**	**£14.61**	**£14.52**	**£13.95**

Section 3

Nutritional Results

National Averages

In this section of the Report, the nutritional value of the food brought into the home throughout 1992 is summarised and compared with the values recorded in 1990 and 1991. In addition there is, for the first time, an assessment of the nutritional contribution made by soft and alcoholic drinks and confectionery brought into the home. More details of the results for 1992 are given in Appendix Tables B9 - B12, where nutrient intakes are also shown for households in different regions and income groups, and with different household compositions. For each category of household, intakes are given not only in absolute amounts per person per day but are also compared with the reference nutrient intakes published by the Department of Health (DH) in 1991[1]. In addition Appendix Table B13 shows the contributions made by selected foods to average nutrient intakes. A more detailed discussion of the fats in the diet is included in Section 4; longer-term trends in the nutritional value of the British diet from 1940 to 1990 were given in the Annual Report for 1990[2].

Energy

The energy content of food brought into the home declined steadily between 1970 and 1990, but now appears to have stabilised. The average household diet in 1992 provided 1,860 kcal per person per day compared with 1,840 kcal in 1991 and 1,870 kcal in 1990. The earlier decline was partly because less food was needed by an increasingly sedentary population and partly because food eaten outside the home and alcoholic drinks and confectionery, which were not recorded at the time, were making an increasing contribution to total energy needs.

Since the beginning of 1992, however, the amounts of soft and alcoholic drinks and confectionery brought home have been recorded by households in the Survey, and their nutritional value has been estimated in the same way as that of the food brought home (see Appendix A). Table B9 is therefore able to show the nutrients in *all* the food and drink brought home as well as those in the household food supply. The inclusion of soft and alcoholic drinks and confectionery brought home raised the calculated energy value from 1,860 kcal to 1,960 kcal per person per day, increasing the calculated proportion of the DH Estimated Average Requirement for energy[1] that was

---

1 Department of Health, *Dietary Reference Values for Food Energy and Nutrients for the United Kingdom,* Report on Health and Social Subjects No.41, HMSO, 1991.
2 Ministry of Agriculture, Fisheries and Food, *Household Food Consumption and Expenditure 1990*, HMSO, 1991.

met from 89 per cent to 93 per cent (after allowance for wastage and meals eaten outside the home).

Of this additional calculated intake of energy, more was derived from soft drinks (44 kcal per person per day) than from alcoholic drinks (24 kcal) or confectionery (32 kcal). The difference between the latter values and the energy content of the total national supplies of alcoholic drinks and confectionery quoted in previous Survey reports is largely accounted for by the amounts of these items eaten or drunk outside the home.

There have been changes in the contributions made to energy intakes by the main groups of foods (Table 3.1). Between 1991 and 1992, there were small increases in the contributions from milk products (with increases from reduced fat milks and yoghurt more than offsetting a decrease from whole milk), and from fish and fish products, vegetables and cereal products. In contrast, there were declines in the contributions made by meat and meat products (mainly by carcase meats), eggs, fats (with decreases from margarine and butter more than offsetting increases from vegetable oils and reduced fat spreads), and by sugar and fruit. These changes are in the directions generally recommended for improvements in health.

Table 3.1
Contributions made by groups of foods to household energy intakes in selected years

kcal per person per day

	1982	1991	1992
Milk and milk products	244	187	196
Cheese	59	63	60
Meat and meat products	354	290	283
Fish	27	22	29
Eggs	38	24	22
Fats	338	253	248
Sugar and preserves	186	113	105
Vegetables	190	182	192
Fruit	60	75	72
Cereals	634	585	599
Other foods	42	51	52
Total	**2,172**	**1,844**	**1,859**

Fats, carbohydrates and fibre

The energy value of food is primarily derived from its fat, protein and carbohydrate (starch and sugars) content. The Department of Health's report on Dietary Reference Values (DRVs)[1] set out changes which should be made in the relative contributions of fats and carbohydrates to improve the health of the nation. Although the total fat content of the diet did not change between 1990 and 1992, the intake of saturated fatty acids fell and that of polyunsaturated fatty acids rose. Table 3.2 shows the improvements in the contributions made by fat and saturates to the energy value of the household diet (excluding alcoholic drinks and confectionery) between 1982 and 1992 and compares them with the DRVs. The Government's White Paper on The Health of the Nation[3] has set the target date of 2005 for the achievement of the recommendations for fat and saturated fatty acids.

Table 3.2
Proportions of household food energy[(a)] derived from fats and carbohydrate

percentage of food energy

	1982	1990	1991	1992	Dietary Reference Values
Fat	42.6	41.6	41.4	41.7	35
of which: saturated fatty acids	18.4	16.6	16.4	16.3	11
Carbohydrate	44.5	44.9	45.3	44.8	50

(a) excluding soft and alcoholic drinks and confectionery

When the contributions made by household purchases of alcoholic and soft drinks and of confectionery are included, the proportions of energy derived from fat, saturated fatty acids and carbohydrate in 1992 change to 40.1, 15.8 and 45.9 per cent respectively. A more detailed assessment of the fat in the diet is included in Section 4.

The total household intake of carbohydrate in 1992 consisted of 53g of non-milk extrinsic sugars and 169g of starch, intrinsic sugars and milk sugar (lactose), compared with 58g and 165g respectively in 1991. This increase in starch and decrease in sugars is also in the direction recommended by the Department of Health. Additionally there are sugars in most soft and alcoholic drinks and in confectionery, and all of them are extrinsic. Household purchases of these items added a further 17g of non-milk extrinsic sugars in 1992 (see Table B9).

The dietary reference value for fibre is 18g per person per day expressed as non-starch polysaccharide, except for young children. The amount in the British household diet was unchanged between 1991 and 1992 at 12g per person per day.

Minerals and vitamins

The amounts of six minerals and nine vitamins in the average household diet in 1992 are compared with intakes in 1990 and 1991 in Appendix Table B9. A further column shows the totals together with the contributions from soft and alcoholic drinks and confectionery. In each case the amounts are compared with the appropriate Reference Nutrient Intakes[1].

There were only small changes in the intake of most of these nutrients between 1991 and 1992. Intakes of calcium rose, almost entirely because of the increase in milk consumption. The intake of vitamin A also

---

3 Department of Health, *The Health of the Nation*, HMSO, 1992.

rose, mainly because an increased contribution from retinol in liver more than offset a decline in β-carotene from carrots and other vegetables. In contrast, intakes of vitamin C fell from 55 mg to 51 mg per day because of declines in both vegetables and fruit (including fruit juices), and vitamin D fell with declining contributions from a number of fatty foods including fortified margarine.

The contributions made by alcoholic and soft drinks and by confectionery to mineral and vitamin intakes are shown for the first time in Table B9. They provided small but useful amounts of iron, magnesium, most B-vitamins, and vitamin C. Total intakes of the vitamins remained well above the Department of Health's recommendations, but intakes of iron, zinc, magnesium and potassium provided less than the RNIs, as they did in 1991.

Regional, Income Group and Household Composition Differences

Nutrient intakes in households in different regions and income groups, and with different family compositions, are shown for 1992 in Appendix Tables B10 - B12. These do not include the contributions from soft and alcoholic drinks or from confectionery. As in previous years the nutritional variations were smaller than the variations in dietary patterns because foods of broadly similar nutritional value, such as beef and lamb or butter and margarine, tend to be substituted for one another.

Intakes of food energy, and therefore of a number of nutrients, were slightly lower in the West Midlands and slightly higher in Yorkshire and Humberside and the East Midlands than elsewhere (Table B10). In most previous years, there has been marked regional variation in the intake of the nutrients derived from fruit and vegetables, with the lowest intakes being in Scotland and the highest in South East England but these were less marked in 1992 as indicated in Table 3.3. The proportion of energy from fat and from saturated fatty acids continued to show little regional variation. There were only small regional differences in the contribution made by alcoholic and soft drinks and confectionery to energy intakes.

Table 3.3
Intakes of vitamin C, β-carotene and folate in selected regions

per person per day

	Vitamin C mg		β-Carotene µg		Folate µg	
	1991	1992	1991	1992	1991	1992
Scotland	46	50	1,460	1,670	209	230
Wales	48	48	1,960	1,870	240	244
England	56	51	1,960	1,750	242	244

Differences in nutrient intakes between households in different income groups are shown in Table B11. Lower income households had a greater energy intake from household food than did higher income households, but a lower intake from soft and alcoholic drinks and confectionery. There was, however, little difference in the intakes of most minerals and vitamins, apart from vitamin C, between income groups A and D.

As in previous years, differences in nutrient intakes varied more with the composition of the household, as shown in Table B12. Households with children had a lower intake *per person* of most nutrients than did households without children, and in households with children, intakes were generally lower the greater was the number of children. This reflects the smaller needs of children, but there was an exception in that intakes in households with four or more children were higher than in some smaller families.

Section 4

Special Analysis - Fats

This Section examines the importance of fats in the diet, with reference first to fats purchased as such for household consumption (cooking and spreading fats) and then to the total fat content of the diet. The analysis mainly covers the period 1975 to 1992.

Cooking and Spreading Fats

Fats, purchased as such (cooking and spreading fats), are one of the largest sources of fat in the diet and, since their acquisition can be recorded directly, the most easily measurable. The first part of this Section examines how consumption of these fats, both in total and by type, has changed over time, and looks at how these patterns vary between households with different characteristics.

National averages

Figure 4.1 shows how consumption of different types of fats, averaged over all household types, has changed over the period from 1975 to 1992. Total consumption of fats has generally declined during that period, from an average of 11.1 ounces per person per week in 1975 to 8.7 ounces in 1992. Within this total, consumption of some types of fats has declined more sharply while other types have increased. The most striking change is the reduction in consumption of butter from 5.6 ounces per person per week in 1975 to only a quarter of that level by 1992. Until the early 1980s, this fall was largely compensated for by increased consumption of soft margarine. Subsequently the continuing

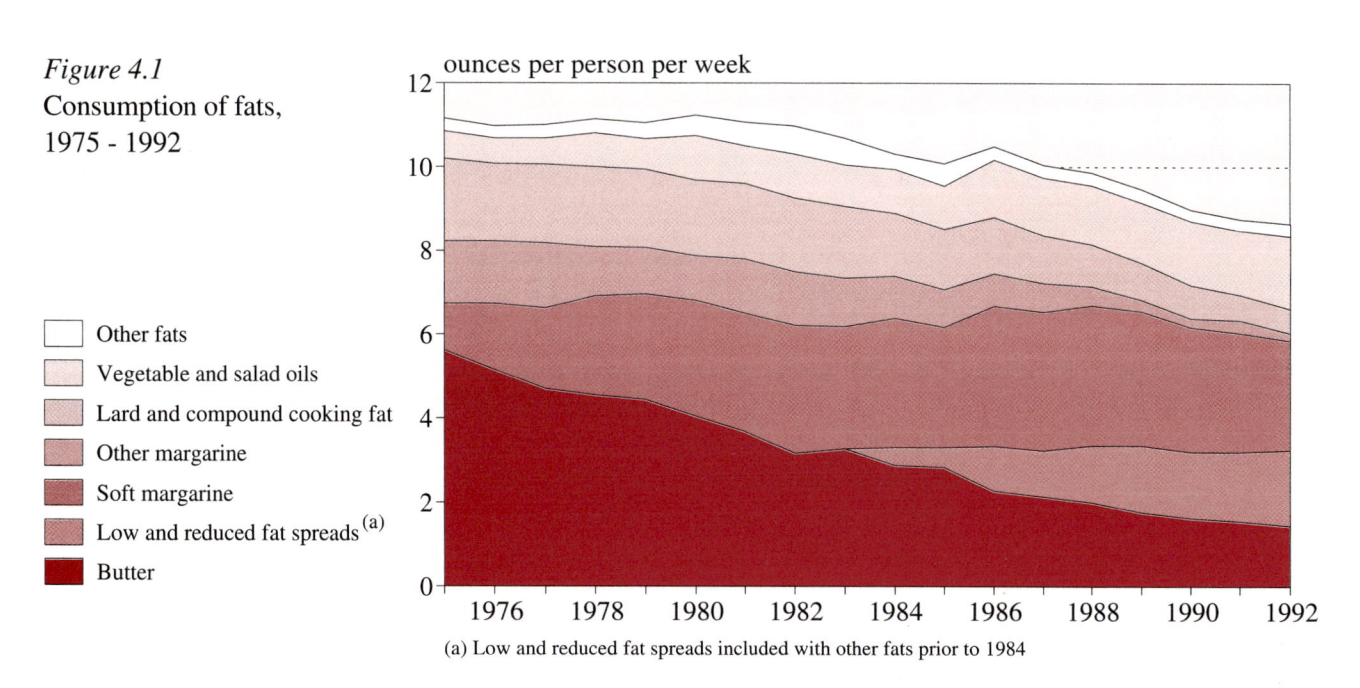

Figure 4.1
Consumption of fats, 1975 - 1992

- Other fats
- Vegetable and salad oils
- Lard and compound cooking fat
- Other margarine
- Soft margarine
- Low and reduced fat spreads [a]
- Butter

(a) Low and reduced fat spreads included with other fats prior to 1984 decline for butter was closely matched by rising consumption of low fat spreads. While consumption of vegetable oils has increased by a half since 1985, quantities of hard margarine and lard and compound cooking fats have fallen very sharply; this, together with a similar decline in purchases of flour, suggests a trend away from home baking.

These changes in the composition of purchased fats are highlighted in Figures 4.2 and 4.3 which relate to yellow and other fats respectively. Figure 4.2 shows that in 1975 butter accounted for over two-thirds of the total consumption of yellow fats. Similar proportions had pertained since the mid-1960s, except for a brief period in the early 1970s when the ratio of butter to margarine dipped sharply. The proportion of soft margarine rose threefold in the decade from 1975, and by 1985 its consumption was the same as that of butter. By 1992, butter accounted for less than a quarter of total yellow fat consumption, while low and reduced fat spreads formed nearly one third. Hard margarine, which was consumed in greater quantities than soft margarine in 1975, had become a very small element of yellow fats consumption by 1992.

Figure 4.2
Composition of consumption of yellow fats

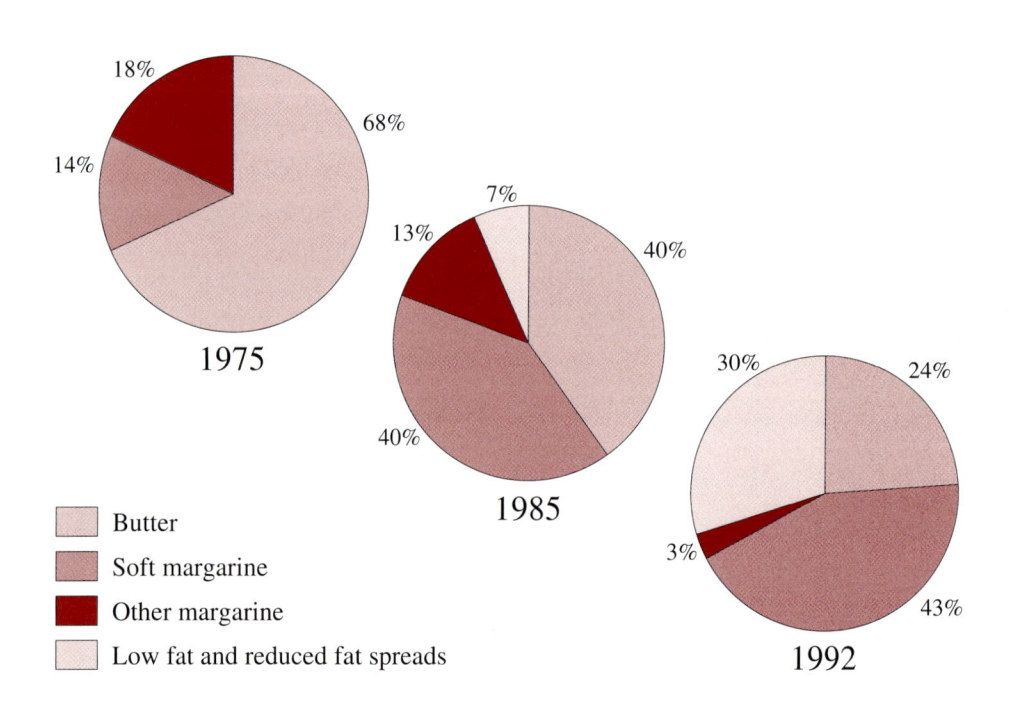

With regard to other fats, Figure 4.3 shows that lard and compound cooking fats comprised over two-thirds of household consumption of other fats in 1975, while vegetable and salad oils comprised less than one quarter. By 1992 the position had reversed, reflecting both the trend away from home baking and increasing consumer concern about saturated fats. The relatively low residual consumption of other fat products includes suet, dripping and fat-based toppings such as cream alternatives.

Figure 4.3
Composition of
consumption of
other fats

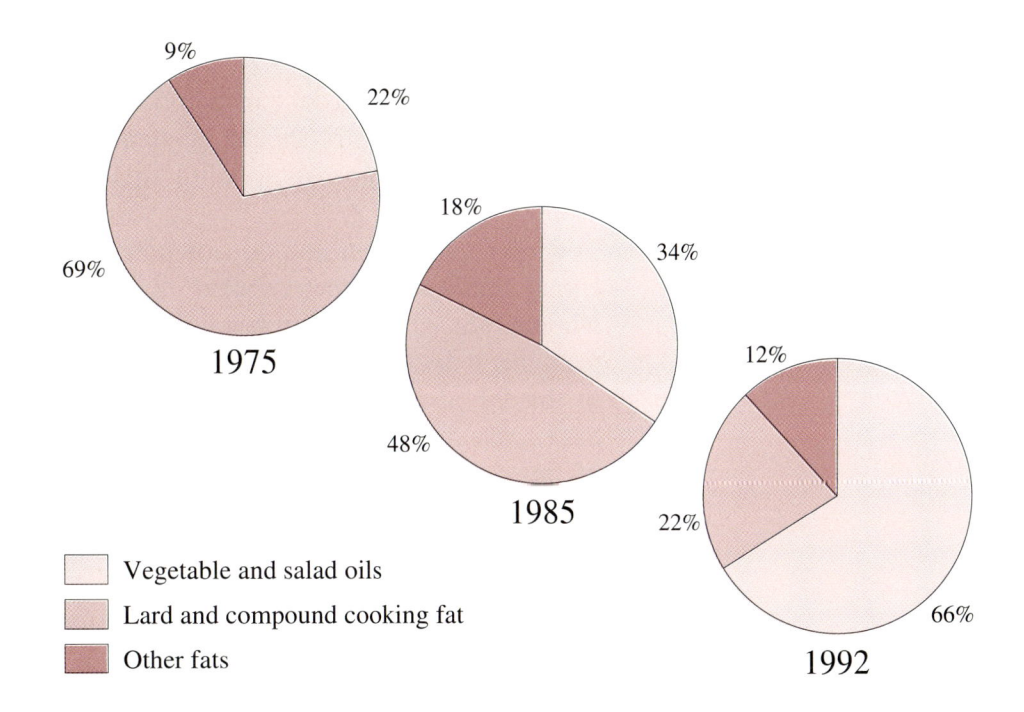

1975: 9% Other fats, 22% Vegetable and salad oils, 69% Lard and compound cooking fat

1985: 18% Other fats, 34% Vegetable and salad oils, 48% Lard and compound cooking fat

1992: 12% Other fats, 66% Vegetable and salad oils, 22% Lard and compound cooking fat

- Vegetable and salad oils
- Lard and compound cooking fat
- Other fats

Regional comparisons

These national changes are largely reflected for each individual region, with only a few noticeable differences. For example, households in Scotland and Yorkshire and Humberside had the lowest average consumption of butter in 1975; however subsequent reductions were less marked there than in other regions and in 1992 they had the highest butter consumption. Wales and the South West, which in 1975 had the highest consumption of butter, have in recent years had the highest consumption of low fat spreads.

Income group comparisons

Although comparisons of consumption by households in different income groups between years should be treated with caution (see Section 2), they do show some interesting differences with regard to fats.

Total per capita consumption of purchased fats in households with an earner tends to increase, rather than decrease, as head of household income falls. This pattern has largely been maintained over time despite the general decline in consumption (Figure 4.4). Consumption is highest in the E1 and OAP households, for which the reduction since 1975 has been less than in any of the other income groups. In 1992 consumption of fats for each of these two groups was over 30 per cent above the average for all households. Broadly similar patterns are found for bread and flour over the same period. The higher fats consumption in E1 and OAP groups may therefore reflect a greater extent of home cooking and baking as well as fewer meals being eaten outside the home by older people.

Figure 4.4
Consumption of fats by income group

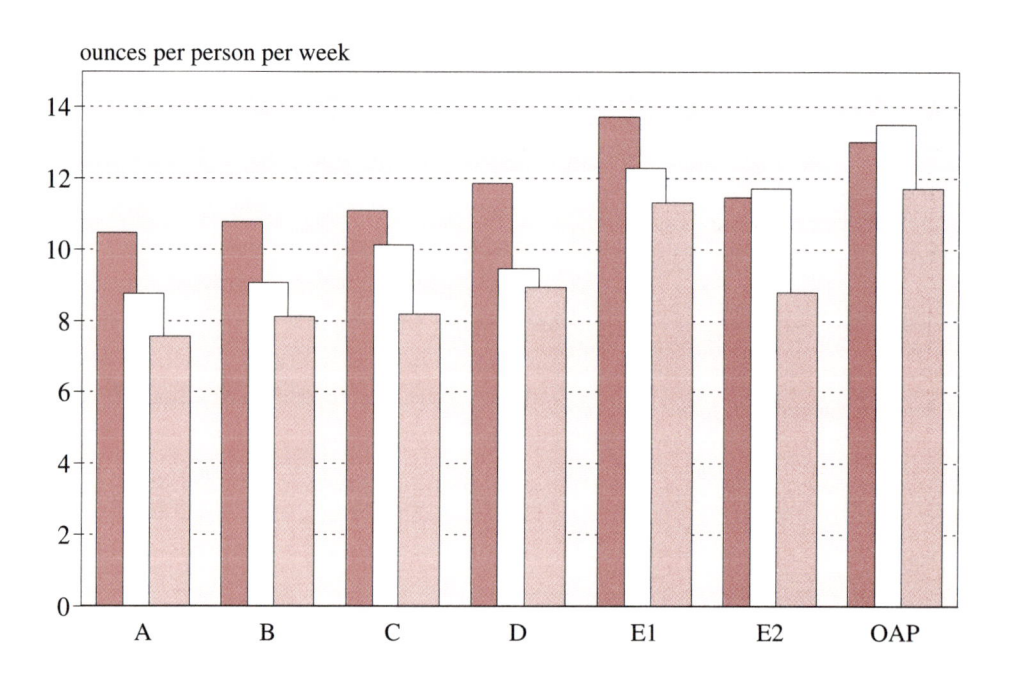

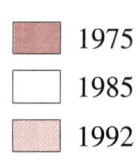

■ 1975
□ 1985
■ 1992

The marked decline in butter consumption since the mid 1970s is observed in most earner groups. Consumption by households in group A has dropped least, and some differences between other groups have become more marked (Figure 4.5). OAP households and those in group E1 had the highest butter consumption in 1975 and, despite a significant decline in their usage of butter, consumption in 1992 for these groups was, respectively, 90 per cent and 60 per cent above the average for all households compared with about 25 per cent above average in 1975.

Figure 4.5
Relative consumption of butter by income group

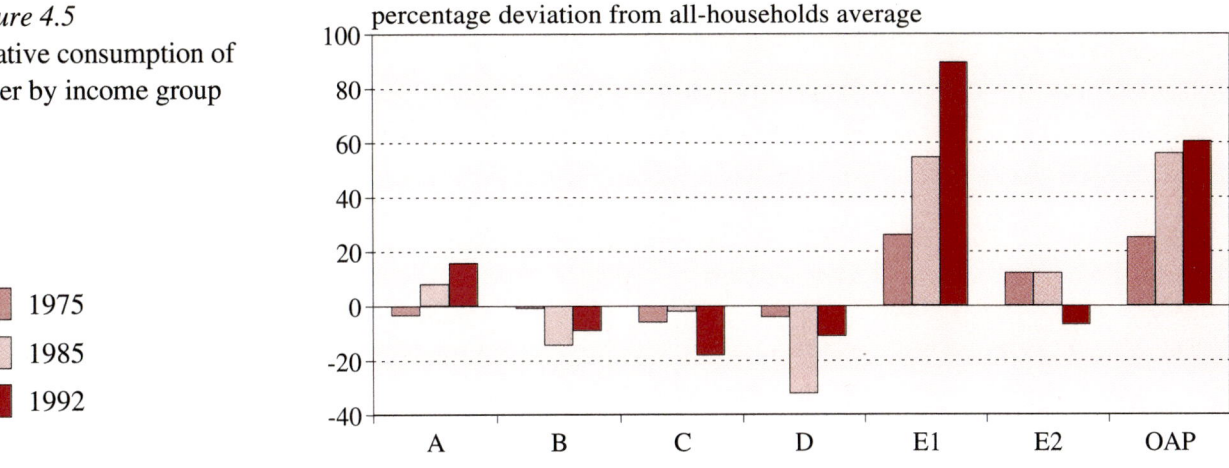

■ 1975
■ 1985
■ 1992

The reduction in consumption of hard margarine is least for households in group E1 and, for these and OAP households, consumption in 1992 was more than twice the average (Figure 4.6). For households with at least one earner, the strong decline in hard margarine consumption with increasing income (of head of household) observed in 1975 and 1985 was negligible in 1992.

Figure 4.6
Consumption of hard margarine by income group

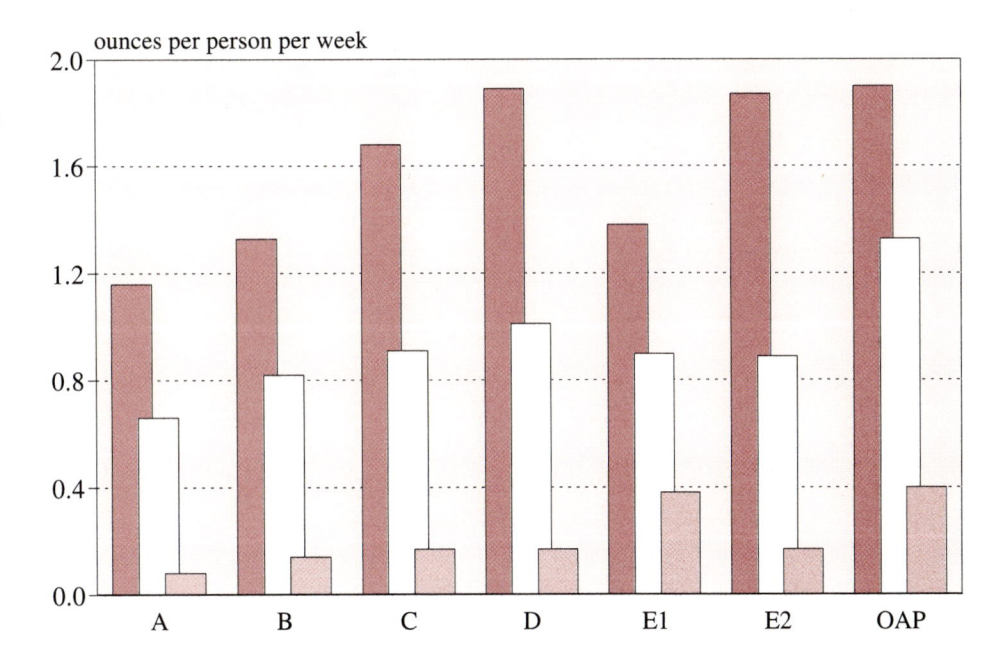

■ 1975
□ 1985
■ 1992

Consumption of soft margarine was markedly higher in lower income groups in 1975 and 1985 (Figure 4.7) but the difference was much less marked in 1992.

Figure 4.7
Consumption of soft margarine by income group

■ 1975
□ 1985
▨ 1992

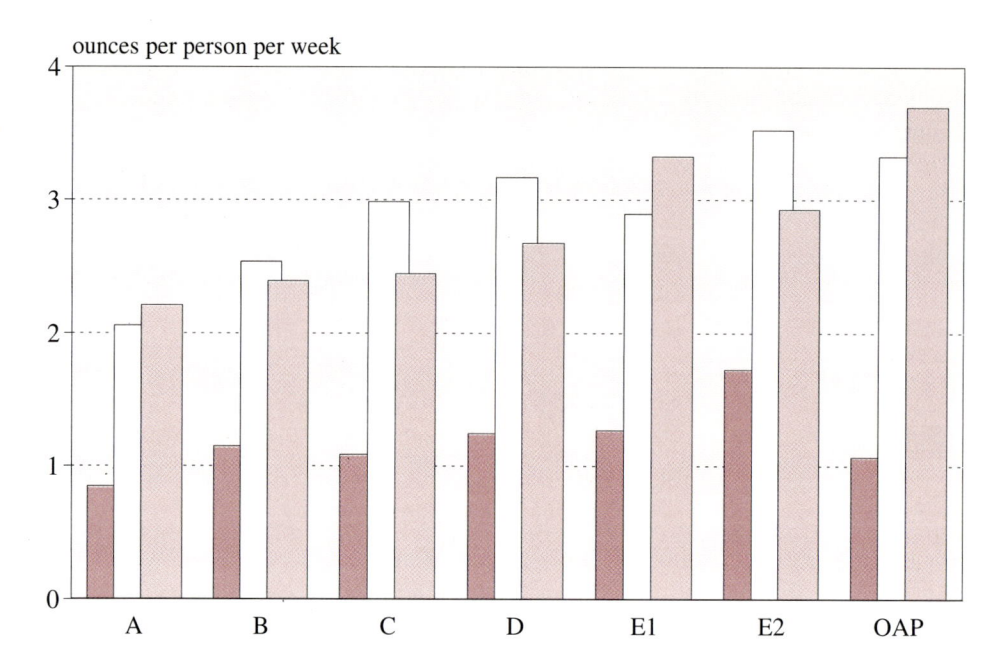

Consumption of vegetable and salad oils used to be generally greater for households in higher income groups (Figure 4.8). In 1992, however, consumption in almost all groups was very similar, the marked exception being OAP households where oils represented only 11 per cent of fats acquisitions compared with an average of 20 per cent for all households.

Figure 4.8
Consumption of vegetable and salad oils by income group

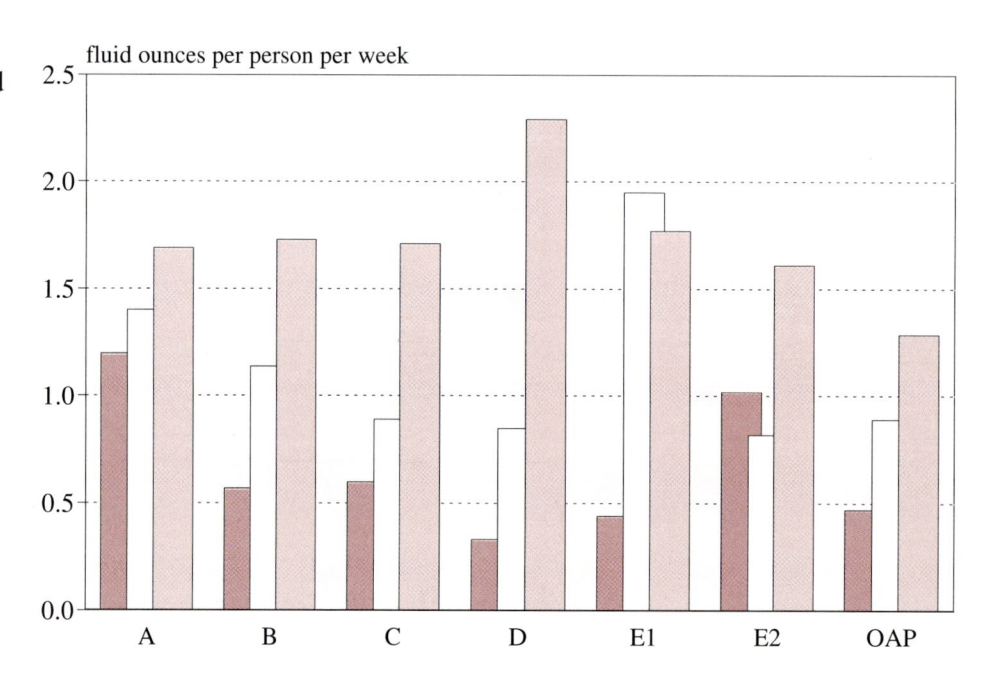

■ 1975
□ 1985
▨ 1992

Household composition comparisons

The overall pattern of consumption of fats by household composition is strongly influenced by whether or not there are children in the household. For each of the three years illustrated, the level of per capita consumption for households consisting of 1 or 2 adults only is about 20 per cent higher than the all-households average; for larger all-adult households it is about 10 per cent higher; and for almost all types of households with children consumption is below the average (Figure 4.9). In 1975 and 1985, average consumption per person was lower for households with larger numbers of children, but in 1992 there was a slight opposite effect.

Figure 4.9
Relative consumption of fats by household composition

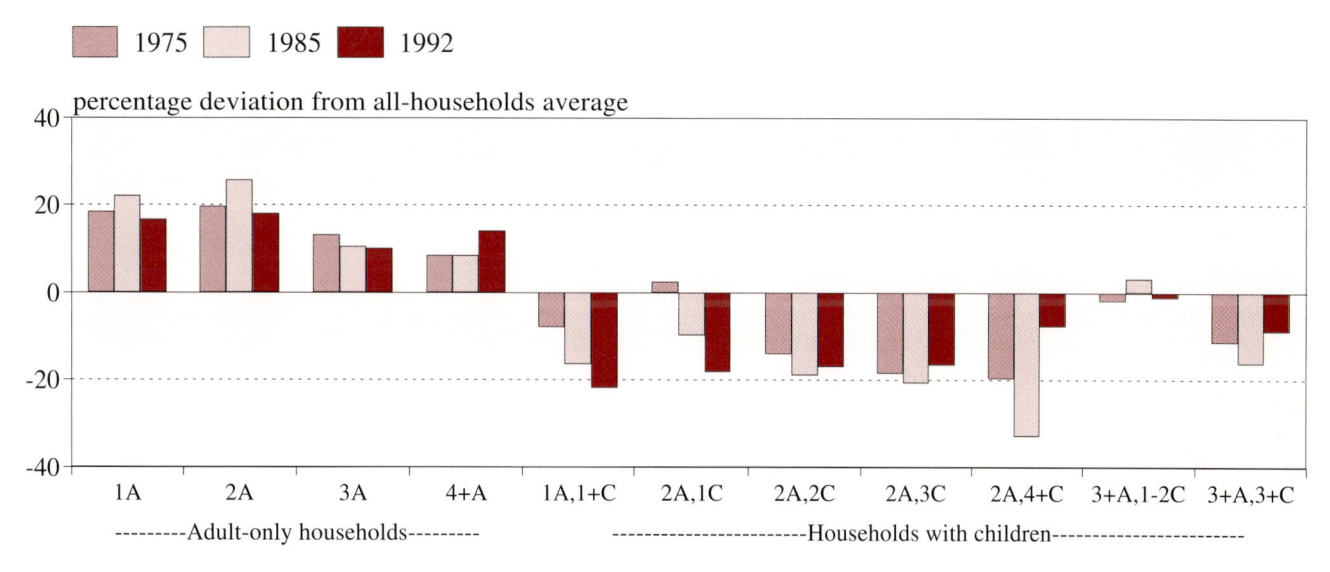

The pattern of consumption of butter is similar to that for all fats, but the difference between adult only households and those with children is more pronounced. In 1975, average consumption of butter ranged from 7.7 ounces per person per week for 1-adult households, to 4.2 ounces for 2-adult households with 3 children. Subsequently, households with children showed a greater reduction in butter consumption than adult-only households, so the differences have become more polarised over time (Figure 4.10).

Figure 4.10
Consumption of butter by household composition

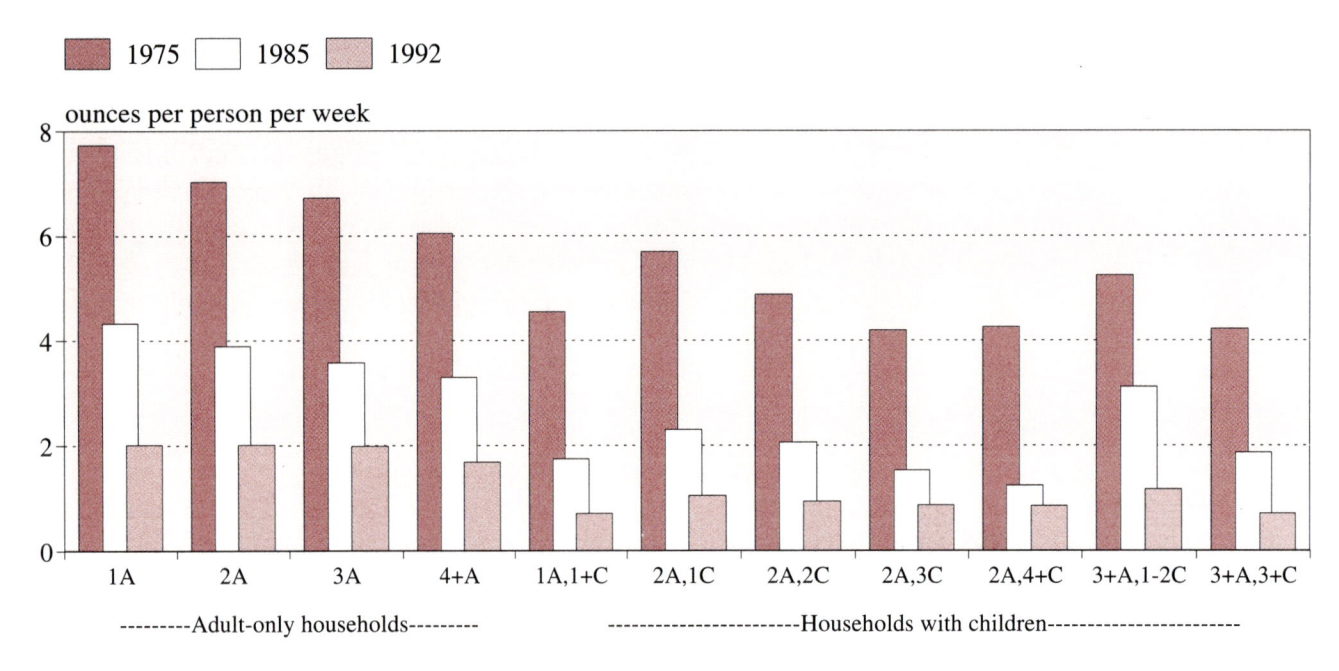

[Bar chart showing ounces per person per week for 1975, 1985, and 1992 across household compositions: 1A, 2A, 3A, 4+A (Adult-only households) and 1A,1+C; 2A,1C; 2A,2C; 2A,3C; 2A,4+C; 3+A,1-2C; 3+A,3+C (Households with children)]

Household composition

Consumption of soft and hard margarine and low or reduced fat spreads was also generally higher in adult-only households than in those with children, but the effect was much less marked than for butter.

Analysis by age of main diary keeper

The main diary-keeper is normally the person most responsible for the household's food purchases and meals and so is likely to have most control over what food items are bought. In this, he or she will be influenced by many factors, including income, household composition (especially the presence of children), and number of meals eaten outside the home. These factors will vary with age, so it is useful to look at the relationship between the age of the diary-keeper and patterns of consumption.

Figure 4.11
Consumption of fats by age of main diary-keeper

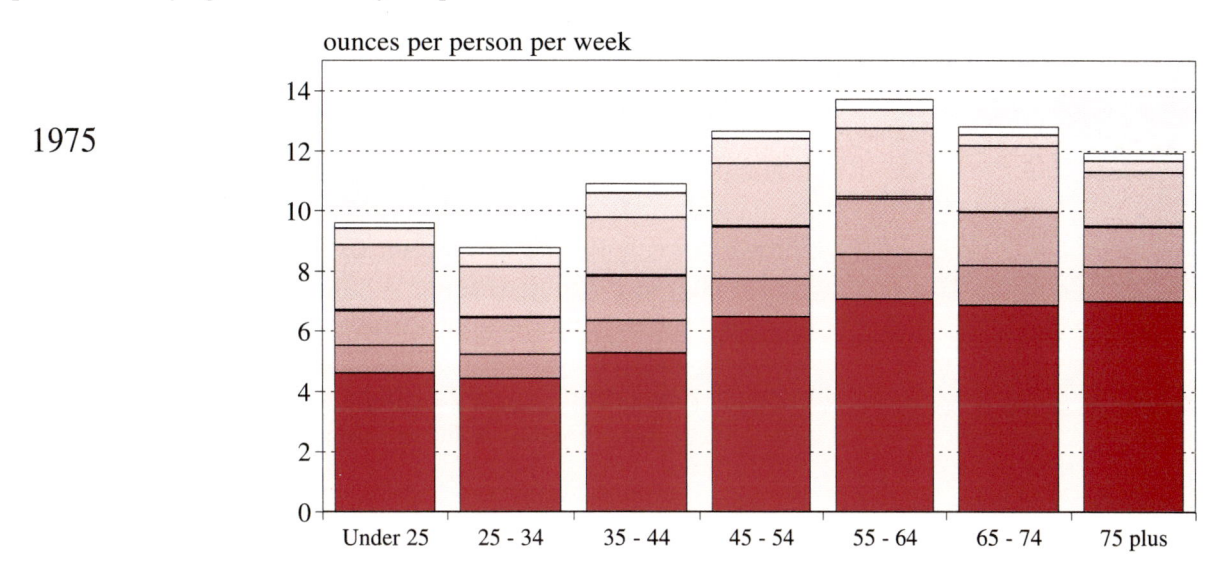

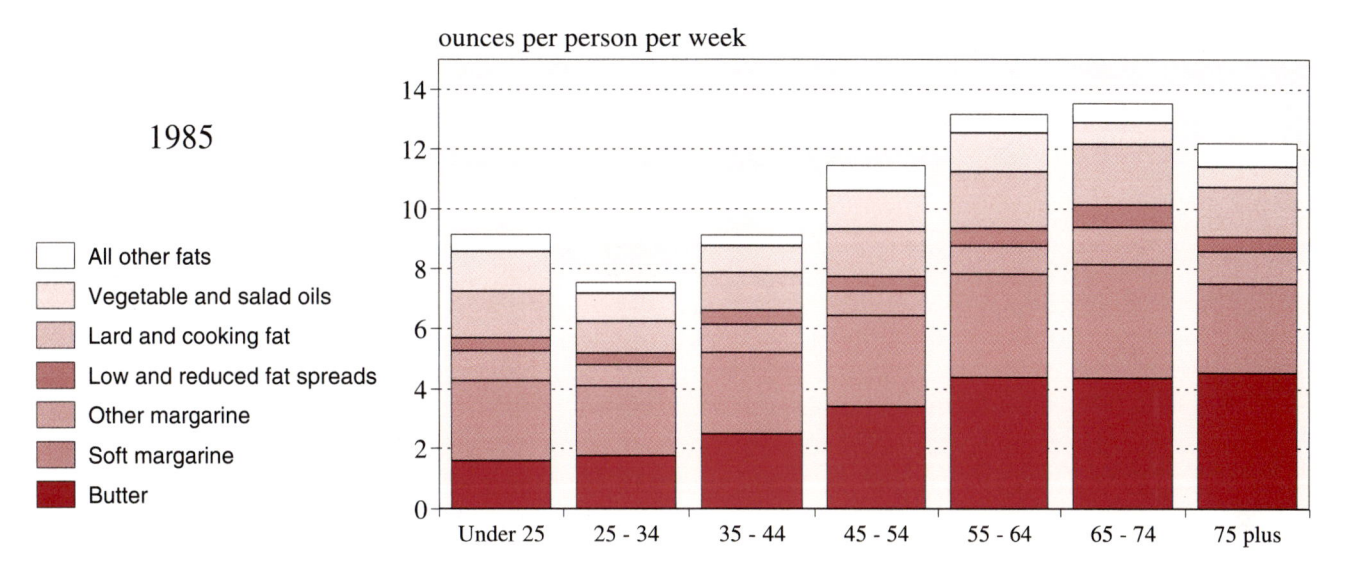

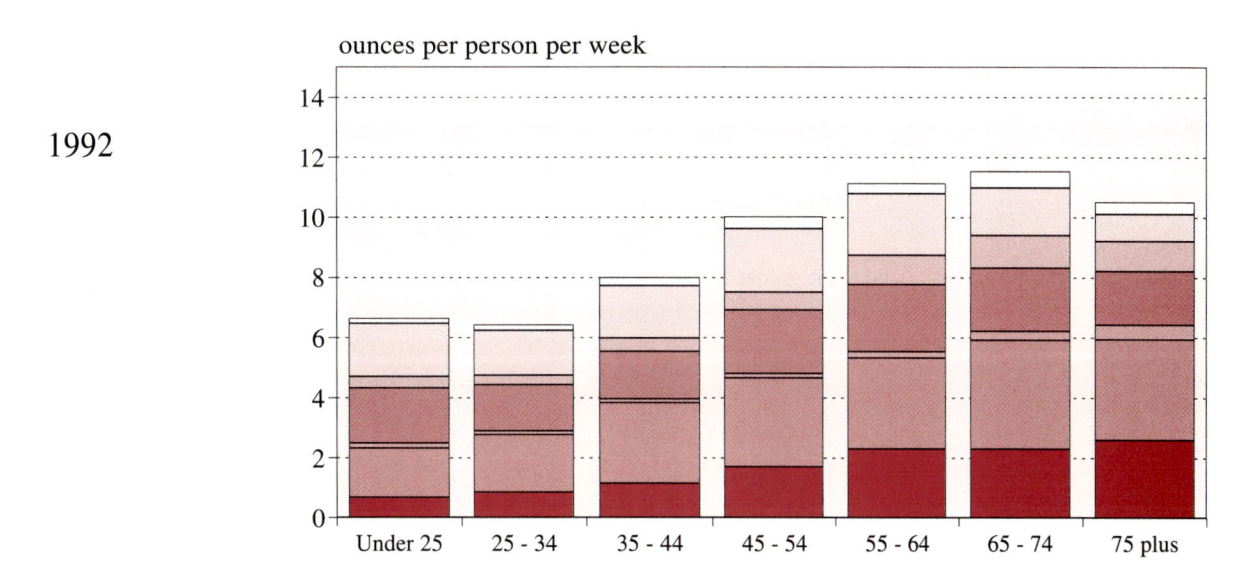

1975

1985

All other fats
Vegetable and salad oils
Lard and cooking fat
Low and reduced fat spreads
Other margarine
Soft margarine
Butter

1992

47

Total consumption of fats was lowest on average in households where the main diary-keeper was aged 25 - 34, then rose to a peak in the 55 - 64 or 64 - 74 age group before declining slightly (Figure 4.11). There was more variation between the age groups in 1992 and 1985 than in 1975.

Although butter consumption also tends to increase with the age of the main diary-keeper, this relationship has changed over time. In 1975 butter formed a similar proportion of total fat purchases in all except the 75 plus age group (Figure 4.12). In 1985, there were marked differences between the age groups, with the proportion of butter increasing from 17 per cent in the under 25 age group to 37 per cent in the 75 plus group. In 1992, although butter consumption was much lower, there was greater relative variation between the age groups.

Figure 4.12
Consumption of butter as percentage of total fats by age group of main diary-keeper

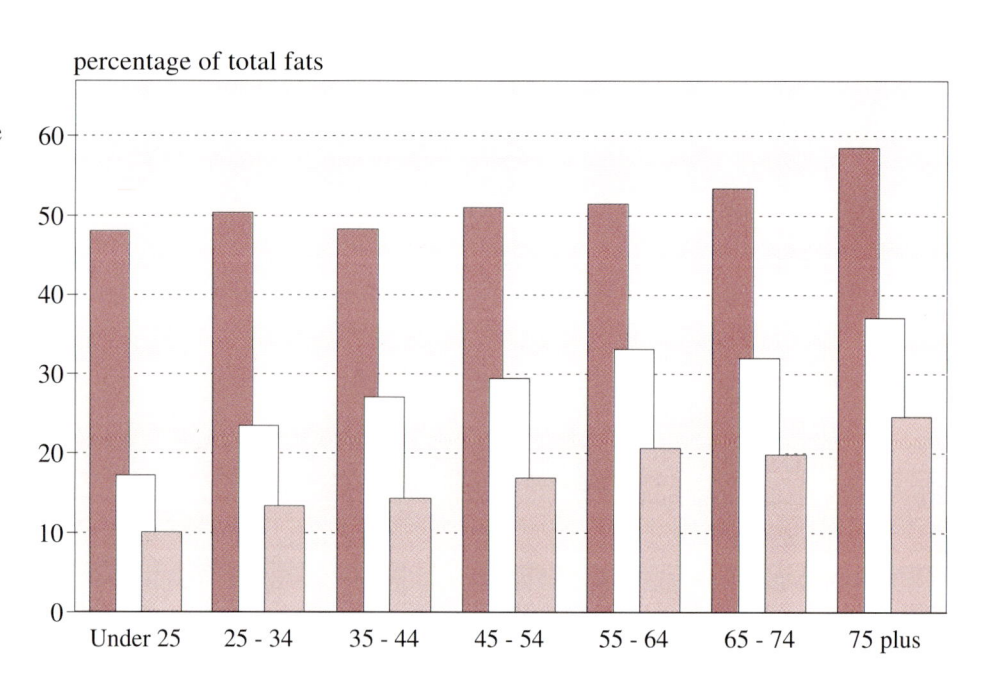

■ 1975
□ 1985
▨ 1992

Consumption of soft margarine also generally increased with the age of the main diary-keeper, with this pattern being more marked in 1992. Hard margarine consumption was highest in the 55 - 64 age group in 1975, but in 1985 and 1992 the consumption rose between successively higher age groups, suggesting a cohort effect. However, the magnitude of the differences were far less marked for soft and hard margarine than for butter. Consumption of low fat spreads in 1992 formed a relatively high proportion of total fats consumption for households with younger diary-keepers, (particularly those under 25). The same was true for vegetable and salad oils, where consumption was relatively low in the two highest age groups.

Total Fat Intakes

Introduction

Almost since the National Food Survey began in 1940, the total fat[1] content of the household diet has been calculated, and the contributions made by the main groups of foods to this total have been presented in the Annual Reports. The proportions of energy derived from fat and from protein and carbohydrate have also been estimated. In addition, since the early 1950s the Survey reports have shown average fat intakes in each of the main regions of Britain and in households in different income groups or with different family compositions. From 1969, the reports have shown the intakes of saturated, monounsaturated and polyunsaturated fatty acids in each category of household, and additional information on individual fatty acids and on the cholesterol content of the household diet has been determined on occasion and published separately. Some of this information is summarised in the following paragraphs.

National average intakes of fat and fatty acids

The total amount of fat in the diet is derived not only from the cooking and spreading fats described in the first part of this Section, but also from a wide range of other fat-containing foods such as dairy products, meats and meat products, and pastry, cakes and biscuits. Fat is also present in some of the items that were were not recorded in the Survey prior to 1992, such as chocolate confectionery. Figure 4.13 shows the total intake of fat in the food brought home since 1943. After the low point of 82g per person per day in 1947, which resulted from the severe postwar food shortages, intakes rose fairly steadily until 1969 when they reached their highest point of 120g per day. They have subsequently declined more quickly, and in 1992 intakes averaged 86g per day.

---

[1] Fat is an important source of energy but the amount of fat in the diet, and the type of fat, are two of the factors which are thought to be associated with coronary heart disease. The prevalence of heart disease is high in the United Kingdom and since 1974 the Department of Health has made a number of recommendations to reduce the amount of fat and of saturated fatty acids in the British diet. Following the first general recommendations, (Department of Health, *Diet and Coronary Heart Disease*, Report on Health and Social Subjects No.7, HMSO, 1974) the second report on diet and cardiovascular disease in 1984 (Department of Health, *Diet and Cardivascular Disease,* Report on Health and Social Subjects No.8, HMSO, 1984) recommended that no more than 35 per cent of food energy should be derived from fat and no more than 15 per cent from saturated fatty acids (including trans unsaturated fatty acids). It also stated that the ratio of polyunsaturated to saturated fatty acids (the P/S ratio) might in consequence increase to 0.45.

In 1991, the report on Dietary Reference Values (Department of Health, *Dietary Reference Values for Food Energy and Nutrients for the United Kingdom,* Report on Health and Social Subjects No.41, HMSO, 1991) recommended that the fat intake of the population should be reduced so that it provided only 33 per cent of the dietary energy which, given that on average alcohol provides 5 per cent of this energy, translates to 35 per cent of the food energy. This report also recommended a reduction in the average intake of saturated fatty acids to 11 per cent of food energy. The Government policy document *The Health of the Nation* (Department of Health, HMSO, 1992) endorsed these targets and set a date of 2005 for their achievement.

These recommendations contrast with earlier concern that the fat content of the diet should be maintained in order to ensure its quality. The first report of the National Food Survey (Ministry of Food, *The Urban Working-Class Household Diet 1940 to 1949,* HMSO, 1951) noted that the fall in fat during the Second World War had decreased the palatability of the national diet and also necessitated an increase in bulk foods so as to maintain energy intakes. Furthermore nutritional standards for school meals until 1975 required that such meals should on average provide at least 32g of fat.

Figure 4.13
Energy and fat intakes, 1943 - 1992

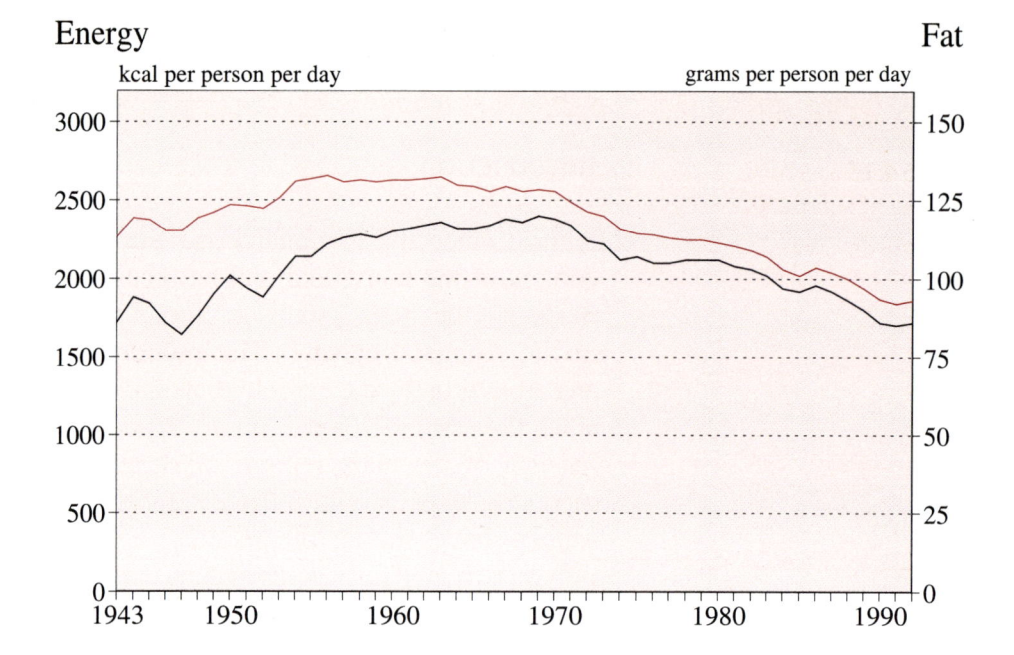

— Energy
— Fat

These changes are largely reflected in total energy intakes from household food, which also rose between the 1940s and the 1960s and then declined steadily until the 1990s (Figure 4.13). The parallel is not exact, however, because there have also been changes in the consumption of carbohydrate-rich foods, especially reductions in bread and potatoes for most of the period, and a marked rise in sugar consumption in the 1950s followed by a steady fall in the 1970s and 1980s. The net result has been that the proportion of food energy derived from fat rose from a low point of 32 per cent in 1947 and has been more than 40 per cent since the mid 1960s (Figure 4.14). It was last at 35 per cent, the target now set by the Department of Health, in 1952, but on average has always been less than the proportion of energy derived from carbohydrate.

Figure 4.14
Percentage of food energy from carbohydrate, fat and protein, 1943 - 1992

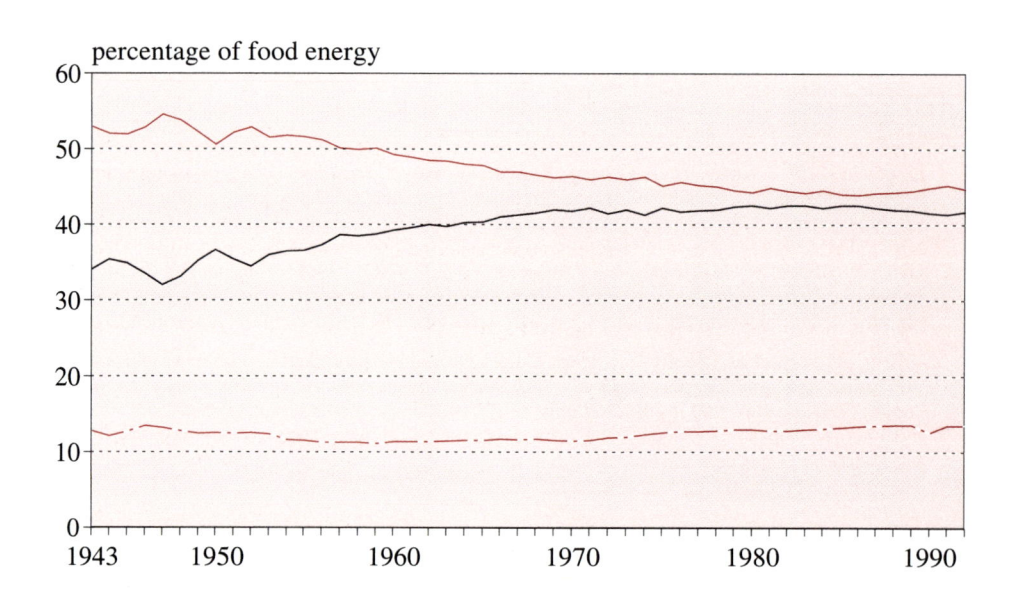

— Carbohydrate
— Fat
—·- Protein

The intakes of saturated, monounsaturated and polyunsaturated fatty acids and the ratio of polyunsaturated to saturated fatty acids (P/S ratio) for selected years since 1975 are shown in Table 4.15.

Table 4.15
Intakes of saturated, monounsaturated and polyunsaturated fatty acids

grams per person per day

	Saturated	Monounsaturated	Polyunsaturated	P/S ratio
1975	51.7	39.8	10.1	0.19
1980	46.8	39.6	11.3	0.24
1985	40.6	34.7	13.1	0.32
1990	34.6	31.8	13.9	0.40
1991	33.7	31.5	13.8	0.41
1992	33.6	31.8	14.4	0.43

Over this period, the intakes of saturated and monunsaturated fatty acids have fallen by 40 per cent and 30 per cent respectively, while the intake of polyunsaturated fatty acids has risen by more than 30 per cent. As a result, the ratio of polyunsaturated to saturated fatty acids has more than doubled to 0.43. However, the target for saturated fatty acid intakes is set in terms of their contribution to energy intakes, and is that they should provide no more than 11 per cent of food energy by the year 2005. The intakes of saturated fatty acids shown in Table 4.15 represented 20.3 per cent of food energy intakes in 1975, but only 16.3 per cent in 1992.

Sources of fat and fatty acids

Throughout the past 50 years, fats purchased as such have been the main contributors to fat in the diet. These have, however, provided only about one third of the total amount of fat in the household diet, with the fat in meat and meat products, dairy products, cereal products and other foods providing the remaining two-thirds. Figure 4.16 shows the relative importance of types of food as purchased to the fat content of the diet since 1975. The contribution of foods *as eaten* will be different, due to some of the fats being used for frying or to make sandwiches, cakes and pastries, etc; furthermore, some visible fat may be discarded, eg from meat, or wasted before or after cooking. The contributions of dairy products and eggs, meat and fish, and fats themselves to total fat have all fallen over this period, while the contributions from cereal products and other foods have risen. These values may be compared with the first such data from the Survey in 1943, when spreading and cooking fats contributed 31g fat per person per day (36 per cent of the total), meats contributed 24g, milk and dairy products 18g and all other foods 13g - values similar to those in the late 1980s.

Figure 4.16
Contributions of foods to fat intakes

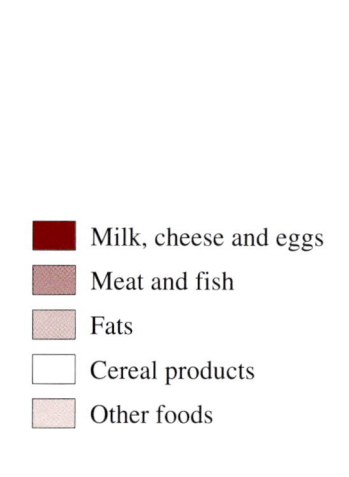

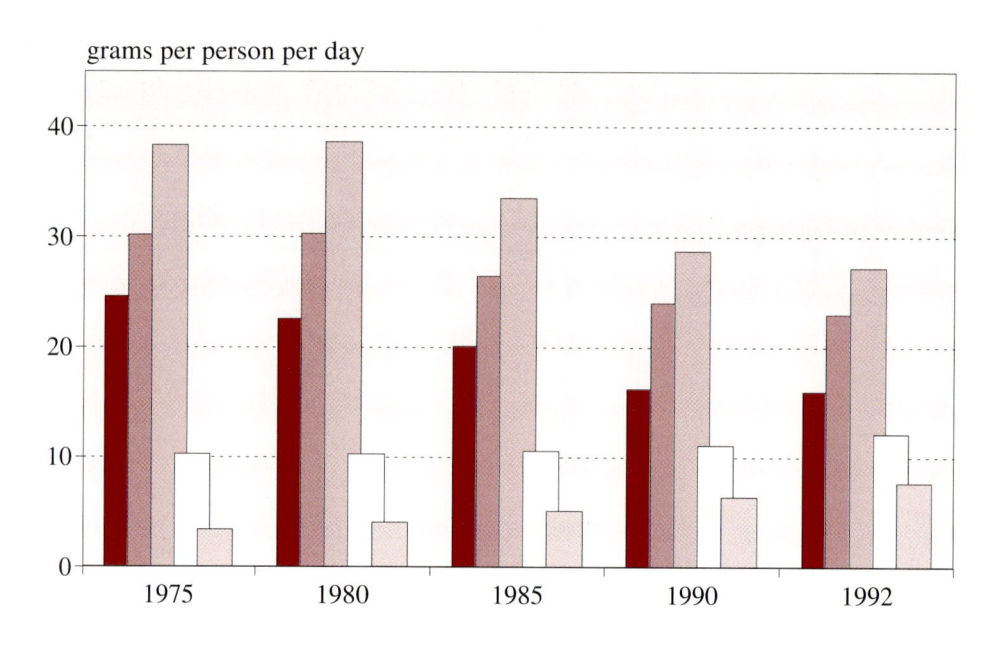

- Milk, cheese and eggs
- Meat and fish
- Fats
- Cereal products
- Other foods

There have been greater changes in the contributors to total fat intakes within these broad groups, mainly reflecting changes in the consumption of the different items but also some changes in their fat content. For example, partly because almost all the milk drunk in 1980 was whole milk (including a significant proportion of Channel Islands milk) whereas more than half is now skimmed or semi-skimmed, the contribution of milk has fallen from an average of 13.4g fat per person per day in 1980 (13 per cent of the total) to only 7.5g (9 per cent) in 1992. On the other hand, the contribution from cheese has risen from 5 to 6 per cent of the total. Within the meats, the contribution of carcase meat, which has been getting leaner, has fallen from 12.7g (12 per cent of the total) to 8.0g (9 per cent). Among the fats, butter, margarine and other fats contributed 13, 12 and 12 per cent of the total fat intake in 1980, but 6, 11 and 12 per cent of a lower total intake in 1992.

Figure 4.17 shows the main contributors to the saturated, monounsaturated and polyunsaturated fatty acids since 1975. Cooking and spreading fats were the main contributors to each, but were especially important sources of the polyunsaturated fatty acids. Meats were also important sources of monounsaturated fatty acids. Meats, dairy products and fats and oils provided approximately equal amounts of saturated fatty acids in the 1990s and for each of these the contribution was smaller than in previous years.

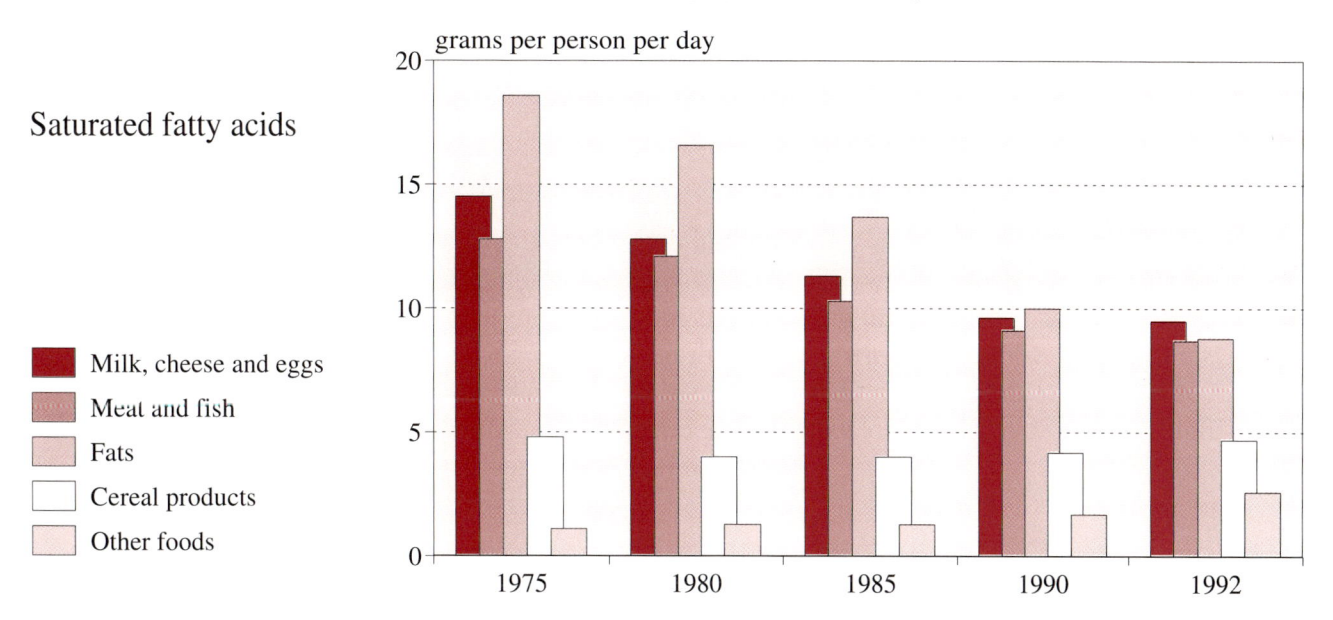

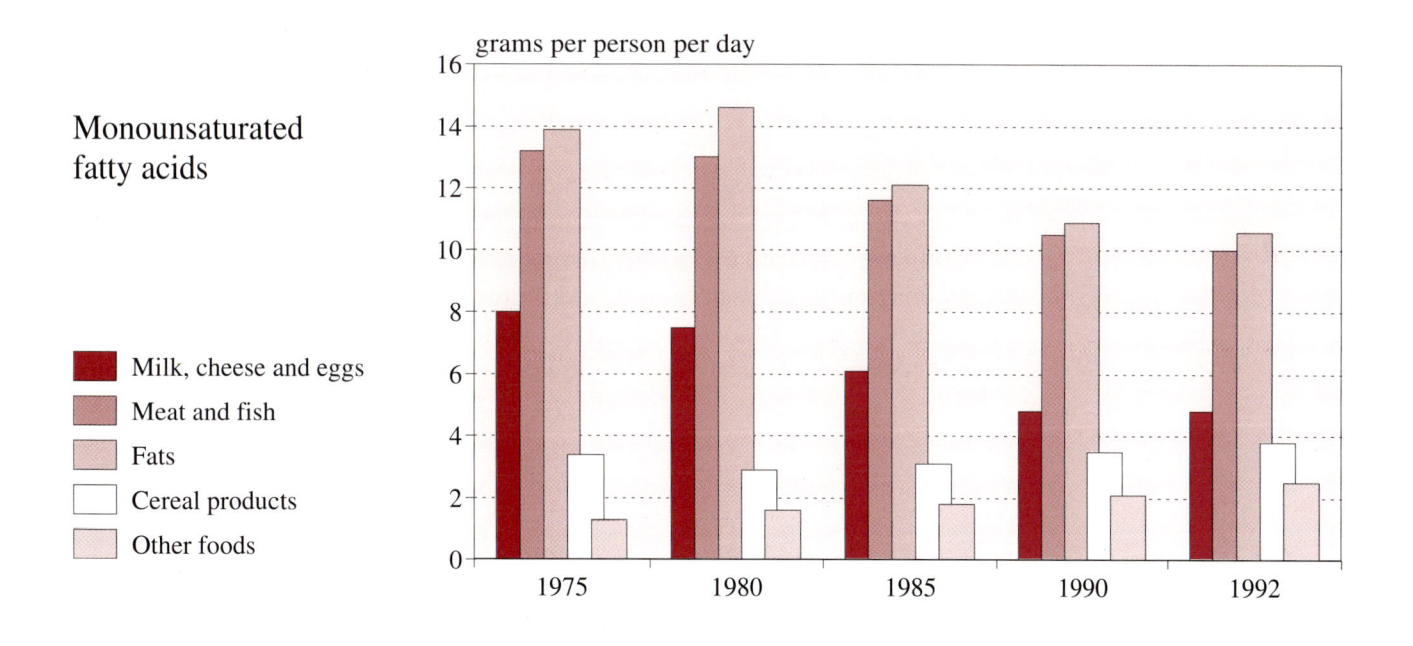

Figure 4.17
Contributions of foods to saturated, monounsaturated and polyunsaturated fatty acid intakes

Saturated fatty acids

- Milk, cheese and eggs
- Meat and fish
- Fats
- Cereal products
- Other foods

Monounsaturated fatty acids

- Milk, cheese and eggs
- Meat and fish
- Fats
- Cereal products
- Other foods

Figure 4.17 continued

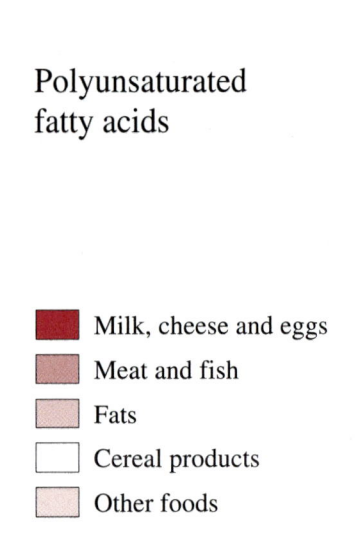

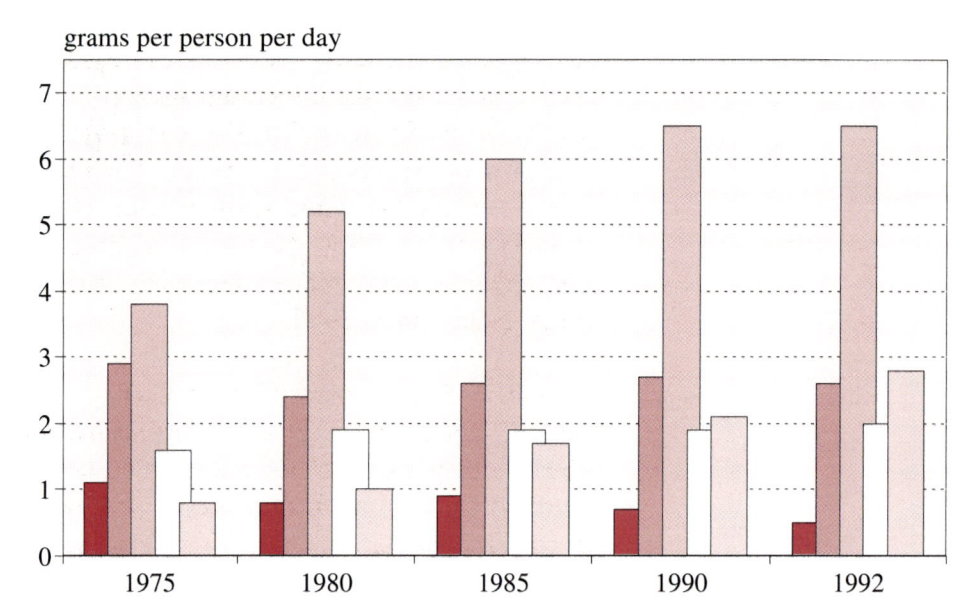

Polyunsaturated fatty acids

- Milk, cheese and eggs
- Meat and fish
- Fats
- Cereal products
- Other foods

grams per person per day

Individual fatty acids and cholesterol

There are many different saturated, monounsaturated and polyunsaturated fatty acids with different properties. The intakes of forty different fatty acids and the main dietary contributors to these intakes were determined from National Food Survey records in 1981, and the results have been published[2]. Household intakes of cholesterol have been estimated twice, with the results showing a fall from 405 mg per person per day in 1970 - 75[3] to 259 mg in 1990[4]. Cholesterol is contained in animal fat, and intakes of animal fats have fallen. The cholesterol contribution from eggs, which are a particularly rich source, fell from 139 mg to 62 mg per day with the decline in egg consumption, and the large decrease in consumption of whole milk and butter also contributed to the reduction in cholesterol. Intakes in England which had been slightly higher than in Scotland are now slightly lower.

---

2 N L Bull, M J L Day, R Burt and D H Buss, *Human Nutrition: Applied Nutrition, 37A,* 1983, pp 373-377.
3 J A Spring, J Robertson and D H Buss, *Proceedings of the Nutrition Society,* 37, 1978, p73A.
4 C McGrath, J Lewis and D H Buss, *Proceedings of the Nutrition Society,* 50, 1991, p231A.

Income group differences

Until 1968, National Food Survey records were analysed according to social class, but since then the analysis has been in terms of the income of the head of household. There is, however, sufficient overlap between the two classifications to permit time trends to be shown, and these are given for the main income bands and for pensioner households in Figure 4.18. Each graph shows the proportions of energy derived from fat and total carbohydrate between 1960 and 1992, and from saturated fatty acids since 1972.

In the 1960s, the proportion of energy derived from carbohydrate was falling, and that from fat was rising, such that in income group A (covering the top 10 per cent of income-earning households) as much energy was derived from fat as from carbohydrate by 1967. Such a point had been reached even earlier, in 1964, in income group A1 (the top 3 per cent of income-earning households), but not until 1971 in group A2. However, because of the stabilisation and subsequent reversal in the relative importance of these two components since then, fat was slightly more important than carbohydrate in income group A until 1986, since when carbohydrate has provided more energy. In income group B the contributions of fat and carbohydrate reached equality at 43 per cent of the energy in 1983; in all other years, however, fat has been less important than carbohydrate, as it has in income groups C and D and in pensioner households throughout the period.

The proportion of energy derived from saturated fatty acids has fallen steadily in each income group, and is now between 16.6 and 16.8 per cent in all of the broad groups shown in the graphs. As with fat itself, saturated fatty acids tended to be more important in the higher income groups, but these income differences in diet have now all but disappeared.

Figure 4.18
Trends in proportions of food energy as fat, carbohydrate and saturated fatty acids by income group

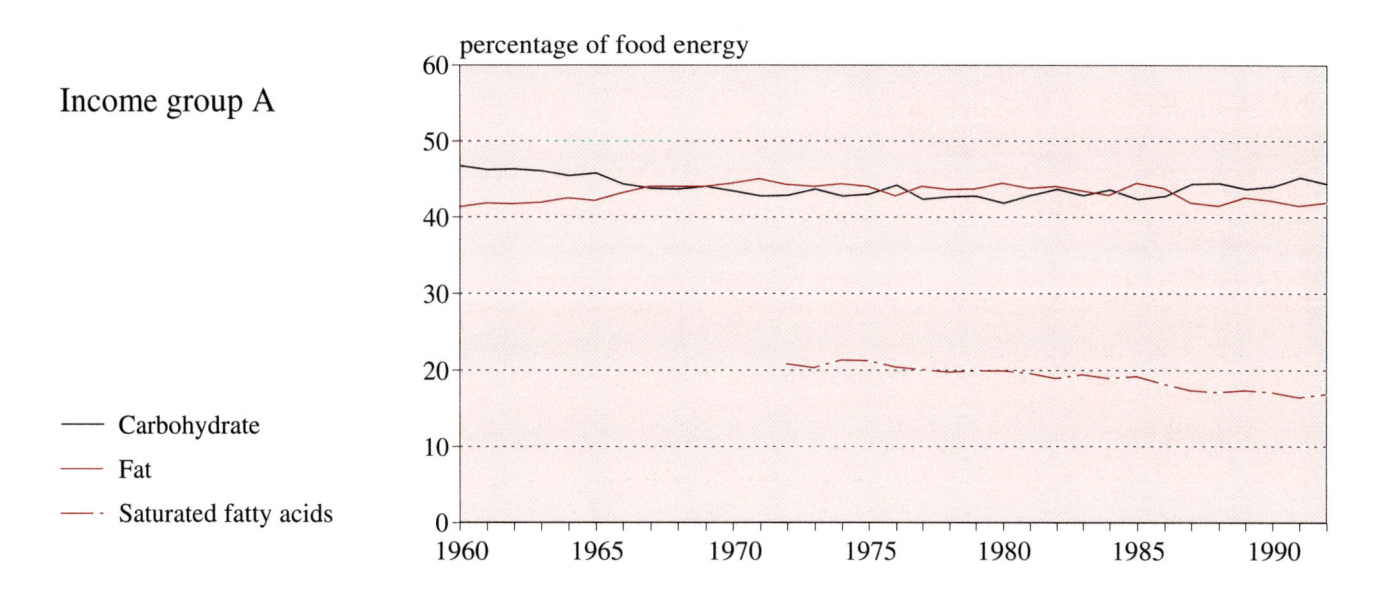

Income group A

— Carbohydrate
— Fat
—· Saturated fatty acids

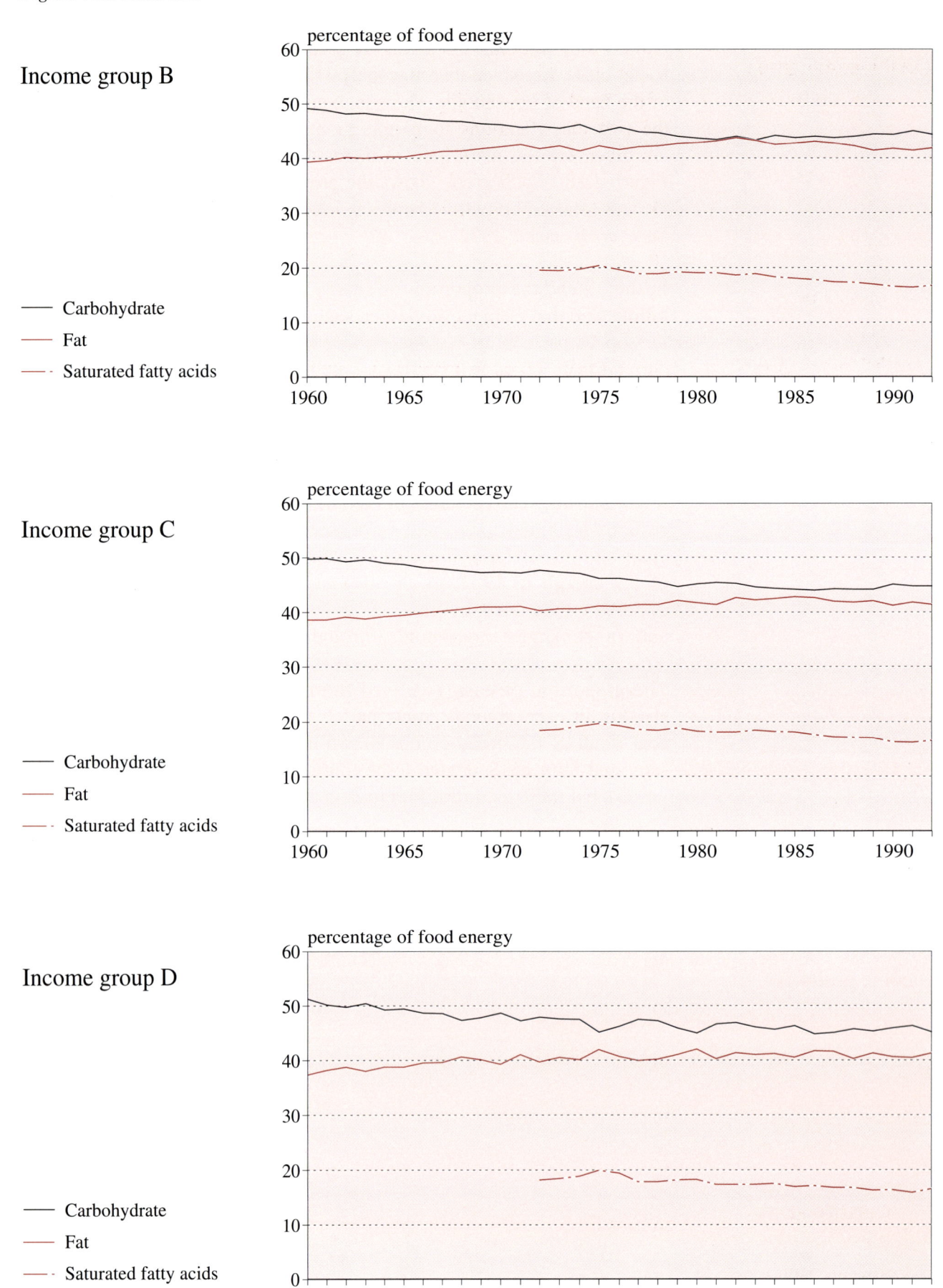

Figure 4.18 continued

Income group B

— Carbohydrate
— Fat
—·— Saturated fatty acids

Income group C

— Carbohydrate
— Fat
—·— Saturated fatty acids

Income group D

— Carbohydrate
— Fat
—·— Saturated fatty acids

Figure 4.18 continued

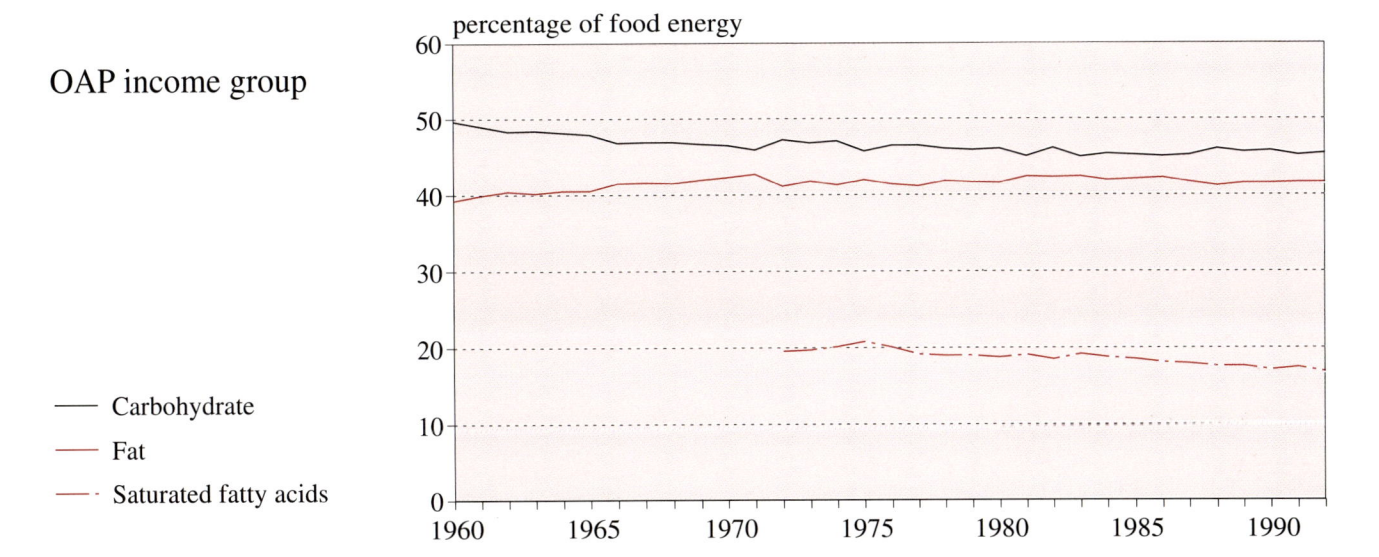

OAP income group

— Carbohydrate
— Fat
—· Saturated fatty acids

Appendix A

Structure of the Survey

Introduction

The National Food Survey is a continuous sampling enquiry into the domestic food consumption and expenditure of private households in Great Britain. Each household which participates does so voluntarily, and without payment, for one week only. By regularly changing the households surveyed, information is obtained continuously throughout the year apart from a short break at Christmas.

Information provided by households

Information from each household is provided by the person, female or male, principally responsible for domestic food arrangements. That person is referred to as the main 'diary-keeper'. She (or he) keeps a record, with guidance from an interviewer, of all the food intended for human consumption entering the home each day. The main Survey[1] therefore excludes meals out (except those based on the household food supply, eg picnics, packed lunches) and pet food.

The following details are noted for each food item: the description, the quantity (in either imperial or metric units) and - in respect of purchases - the cost. Food obtained free from a farm or other business owned by a household member or from a garden or allotment is recorded only at the time it is used. To avoid the double counting of purchases, gifts of food are excluded if they were bought by the donating households. Also now included in the Survey are soft[2] and alcoholic drinks and chocolate and sugar confectionery; these are, however, items which individual household members may purchase for themselves without necessarily coming to the attention of the main diary-keeper. These results are therefore presented separately from the main analyses in this Report because of the possibility of bias arising from their under-recording.

As well as the detail about the foods entering the household, the diary-keeper notes which persons (including visitors) are present at each meal together with a description of the type (but not the quantities) of food served. This enables an approximate check to be made between the foods served and those acquired during the week. Records are also kept of the number and nature (whether lunch, dinner, etc) of the meals obtained outside the home by each member of the household; this is used in the

---

[1] A supplementary survey, conducted alongside the main Survey in 1992, did collect some information on meals out but the results are not presented in this Report.

[2] Particulars of soft drinks bought for the household supply have been collected since 1975.

nutritional calculations - see below. The quantity of school milk consumed by children is also recorded.

On a separate questionnaire, details are entered of the characteristics of the family and its members but names are not collected. The identities of addresses are strictly confidential. They are known only to those who were involved respectively with selecting the sample and carrying out the fieldwork. They are not even divulged to the Ministry of Agriculture, Fisheries and Food which is responsible for analysing and reporting the Survey results.

As the main Survey records only the quantities of food entering the household, and not the amount actually consumed, it cannot provide meaningful frequency distributions of households classified according to levels of food eaten or of nutrition. However, averaged over sufficient households, the quantities recorded should equate with consumption (in the widest possible sense, including waste food discarded or fed to pets) provided purchasing habits are not disturbed and there is no general accumulation or depletion of household food stocks.

The National Food Survey is selected to be representative of mainland Great Britain (including the Isle of Wight but not the Scilly Isles or the islands of Scotland). A three-stage stratified random sampling scheme is used for the selection of addresses, the first stage of which involves the selection of local authority districts as the primary sampling units (P.S.U.s). There are 52 local authority districts included in the Survey each quarter for sampling purposes. Six or seven of the districts are retired and replaced each quarter and those selected then remain in the Survey for eight consecutive quarters (re-selection being possible). The local authority districts surveyed at some time during 1992 are listed in Table A1 of this Appendix. The 52 districts in the Survey in any one quarter are randomly divided into two sets of 26, which are worked alternately for two successive periods of 10 days, there being 32 such periods in the year. In total each selected district is surveyed 30 times during its eight quarters in the Survey, ie 3 or 4 times each quarter.

The second stage of the selection procedure involves the selection of postal sectors within each of the districts and the third is the selection of 18 delivery points from each postal sector. The delivery points are drawn from the Small Users Postcode Address File (PAF) using interval sampling from a random origin.

In 1992, 780 postal sectors were selected at the second stage of sampling and 14,040 addresses at the third stage. When visited, a few of these addresses were found to be those of institutions or other establishments not eligible for inclusion in the Survey; others were unoccupied or had been demolished. In addition, some addresses were found to contain more than one household. After allowing for these factors, the estimated eligible number of households in the sample was 12,309. In some households the diary-keeper was seen but refused to give any information; a number of other diary-keepers answered a questionnaire[3] but declined to keep a week's record; while some who undertook to keep a record did not in fact complete it. Finally, some records were lost in the post or were rejected at the editing stage. The result was a responding sample of 7,556 households, representing 61 per cent of the eligible sample. Details are as follows:-

	Households	Households selected (%)
Number of households at the address selected in the sample	12,309	100
Number that could not be visited for operational reasons	–	–
Number visited but no contact made with the diary-keeper	1,390	11
Interview refused or not practicable	1,593	13
Diary-keeper answered a questionnaire but declined to keep a week's record	866	7
Diary-keeper started to keep a week's record but did not complete it	682	6
Completed records lost in the post or rejected at the editing stage	222	2
Number of responding households	7,556	61

Nutritional analysis of survey results

The energy value and nutrient content of food obtained for consumption in the home[4] are evaluated using special tables for food composition. The nutrient conversion factors are mainly based on values given in *The Composition of Foods*[5] but are thoroughly revised each year for two reasons. First, to reflect changes in nutrient values resulting from new methods of food production, handling and fortification. Second, to reflect changes in the structure of the food categories used in the Survey - for example changes in the relative importance of the many products grouped under 'reduced fat spreads'.

---

3 The questionnaire relates to family composition, occupation, etc
4 See Glossary
5 B Holland, A A Welch, I D Unwin, D H Buss, A A Paul and D A T Southgate, *McCance and Widdowson's The Composition of Foods,* 5th edition, Royal Society of Chemistry and Ministry of Agriculture, Fisheries and Food, Royal Society of Chemistry, 1991

The nutrient factors used make allowance for inedible materials such as the bones in meat and the outer leaves and skins of vegetables. For certain foods, such as potatoes and carrots, allowance is also made for seasonal variations in this wastage and/or nutrient content. Further allowance is made for the expected cooking losses of thiamin and vitamin C; average thiamin retention factors are applied to appropriate items within each major food group and the (weighted) average loss over the *whole* diet is estimated to be about 20 per cent. The losses of vitamin C are set at 75 per cent for green vegetables and 50 per cent for other vegetables. However, no allowance is made for wastage of *edible* food. The exception is when the adequacy of the diet is being assessed in comparison with recommended intakes (see the final paragraph of this Appendix). Then, the assumption is made that in each type of household, 10 per cent of all foods - and hence of all nutrients available for consumption - is either lost through wastage or spoilage in the kitchen or on the plate, or fed to domestic pets[6].

The energy content of the food is calculated from the protein, fat and available carbohydrate (expressed as monosaccharide) contents using the respective conversion factors, (4, 9 and 3.75 kcal per gram). It is expressed both in kilocalories and megajoules (1,000 kcal = 4.184 MJ). Niacin is expressed as niacin equivalent, which includes one-sixtieth of the tryptophan content of the protein in the food. Vitamin A activity is expressed as micrograms of retinol equivalent, that is the sum of the weights of retinol and one-sixth of the β-carotene. Fatty acids are grouped according to the number of double bonds present, that is into saturated, monounsaturated and polyunsaturated fatty acids. For the diet as a whole, the total fatty acids constitute about 95 per cent of the weight of the fat. This proportion varies slightly for individual foods, being lower for dairy fats with their greater content of short-chain acids and a little higher for most other foods.

The nutritional results are tabulated in two main ways for each category of household in the Survey -

(a) *Per person*. This presentation is directly comparable to the per person presentation in Section 2 of this Report of the amounts of food obtained. However, it has some drawbacks where the interpretation of nutrient intakes is concerned. It does not take into account contributions made by meals outside the home or by food outside the diary-keepers' purview (eg confectionery or drinks bought for household consumption without the knowledge of the diary-keeper). Nor is any allowance made for the wastage of edible food. The averages per person can also be misleading. For example, average per capita energy intakes in families with small children are invariably less than those for wholly adult households but

---

6 An enquiry into the amounts of potentially edible food which are thrown away or fed to pets in Great Britain recorded an average wastage of about 6 per cent of households' food supplies (see R W Wenlock, D H Buss, B J Derry and E J Dixon, *British Journal of Nutrition, 43,* 1980, pp 53-70). However, this is considered likely to be a minimum estimate, and the conventional Survey deduction of 10 per cent has been retained thereby preserving continuity with previous years.

this does not by itself indicate that the former are less well nourished because, on average, children have a smaller absolute need for energy.

(b) *As a proportion of Dietary Reference Values published by DH*.[7] Some of the above drawbacks are overcome in this presentation. It involves comparing intakes with household needs after the age, sex and pregnancy of each member have been taken into account. Allowance is also made for meals eaten outside the home and for the presence of visitors by redefining, in effect, the number of people consuming the household food - *not* by adding or subtracting estimates of the nutrient content of the meals in question. Moreover, for these comparisons, the estimated energy and nutrient contents are reduced throughout by 10 per cent to allow for wastage of edible food (see footnote on the previous page). This difference should be borne in mind when comparing the Survey results with the Reference Values.

---

[7] Department of Health. *Dietary Reference Values for Food Energy and Nutrients for the United Kingdom.* Report on Health and Social Subjects No. 41, HMSO, 1991

Table A1
Local Authority Districts surveyed in 1992

Region[a]	Definition of region[a]	Local Authority Districts[b] selected in the sample for 1992
England: North	Cleveland, Cumbria, Durham, Northumberland, Tyne and Wear	*South Tyneside Derwentside Blyth Valley
Yorkshire and Humberside	Humberside, North Yorkshire, South Yorkshire, West Yorkshire, East Yorkshire	*Kirklees *Leeds York *Wakefield Scarborough
North West	Cheshire, Lancashire, Greater Manchester, Merseyside	*Tameside *Sefton Ribble Valley *Manchester *Wirral Chorley Burnley *St Helens
East Midlands	Derbyshire, Leicestershire, Lincolnshire, Northamptonshire, Nottinghamshire	Leicester West Lindsey Erewash
West Midlands	Hereford and Worcester, Shropshire, Staffordshire, Warwickshire, West Midlands	Bromsgrove *Wolverhampton Stoke-on-Trent Cannock Chase Stafford *Dudley Nuneaton Bridgnorth
South West	Avon, Cornwall, Devon, Dorset, Gloucestershire, Somerset, Wiltshire	Restormel Mendip Plymouth Torridge Woodspring
East Anglia	Cambridgeshire, Norfolk, Suffolk	Cambridge South Cambridgeshire
South East	Greater London, Bedfordshire, Berkshire, Buckinghamshire, East Sussex, Essex, Hampshire, Hertfordshire, Isle of Wight, Kent, Oxfordshire, West Sussex	*Barnet *Bromley *Ealing *Wandsworth *Kensington & Chelsea *Hounslow *Camden *Lewisham *Redbridge *Waltham Forest Medina Castle Point Southend Canterbury East Hertfordshire

Table A1 *continued*

Region[a]	Definition of region[a]	Local Authority Districts[b] selected in the sample for 1992
South East (continued)		Cherwell Three Rivers Chichester Winchester Vale of White Horse Reading Stevenage Rochester upon Medway Wycombe Lewes
Wales	The whole of Wales	Colwyn Swansea Alyn and Deeside
Scotland	The whole of Scotland	Dundee *Glasgow *Eastwood Banff and Buchan Edinburgh Badenoch and Strathspey Perth and Kinross Kyle and Carrick

(a) these are the standard regions as revised with effect from 1st April 1974

(b) Local Authority Districts marked * are wholly or partly within Greater London, the Metropolitan districts, or the Central Clydeside conurbation

Table A2
Distribution of the 1992 Survey sample according to income

Income group	Gross weekly income of head of household[a]	Number of households in whole sample		Percentage of households in groups A1 to D	
				realised	target
Households containing one or more earners[b]					
A1	£730 or more	147	1.9	2.9	3
A2	£520 and under £730	333	4.4	6.5	7
B	£280 and under £520	1,950	25.8	38.0	40
C	£140 and under £280	2,023	26.8	39.4	40
D	Less than £140	685	9.1	13.3	10
Total A1 to D		5,138	67.5	100	100
Households without an earner[b]					
E1	£140 or more	461	6.1		
E2	Less than £140	943	12.5		
Pensioner households[c]					
OAP	na	1,014	13.4		
Total all households		7,556	100		

(a) or of the principal earner if the income of the head of household was below £140 (the upper limit for group D)

(b) by convention, the short-term unemployed are classified as 'earners', until they have been out of work for more than a year when unemployment benefit ceases

(c) see Glossary

Table A3
Composition of the sample of responding households, 1992

	Households		Persons		Average number of persons per household	% of households owning a:	
	Number	%	Number	%		deep-freezer	micro-wave
All households	7,556	100	19,348	100	2.56	87	61
Analysis by region							
Scotland	663	8.8	1,668	8.6	2.52	85	62
Wales	457	6.1	1,159	6.0	2.54	88	66
England	6,436	85.4	16,521	85.4	2.57	87	61
North	478	6.3	1,206	6.2	2.52	83	64
Yorkshire and Humberside	547	7.2	1,371	7.1	2.51	84	66
North West	929	12.3	2,399	12.4	2.58	83	60
East Midlands	505	6.7	1,341	6.9	2.66	85	59
West Midlands	750	9.9	1,989	10.3	2.65	86	63
South West	655	8.7	1,677	8.7	2.56	89	62
South East/East Anglia	2,572	34.0	6,538	33.8	2.54	89	59
Analysis by income group [a]							
A1	147	1.9	483	2.5	3.29	95	85
A2	333	4.4	1,055	5.5	3.17	97	80
B	1,950	25.8	5,901	30.5	3.03	93	74
C	2,023	26.8	5,810	30.0	2.87	92	70
D	685	9.1	1,862	9.6	2.72	82	60
E1	461	6.1	864	4.5	1.87	89	54
E2	943	12.5	1,902	9.8	2.02	78	44
OAP	1,014	13.4	1,471	7.6	1.45	70	31
Analysis by household composition [b]							
Number of adults / Number of children							
1 / 0	1,676	22.2	1,676	8.7	1.00	70	40
1 / 1 or more	306	4.1	866	4.5	2.83	83	58
2 / 0	2,517	33.3	5,034	26.0	2.00	89	60
2 / 1	681	9.0	2,043	10.5	3.00	93	72
2 / 2	903	11.9	3,612	18.7	4.00	96	77
2 / 3	289	3.8	1,445	7.5	5.00	98	73
2 / 4 or more	72	0.9	445	2.3	6.18	94	63
3 / 0	550	7.3	1,650	8.5	3.00	93	71
3 or more / 1 or 2	326	4.3	1,481	7.6	4.54	97	78
3 or more / 3 or more	43	0.6	288	1.5	6.70	93	79
4 or more / 0	193	2.6	808	4.2	4.19	95	78
Analysis by ownership of dwelling							
Unfurnished, council	1,453	19.2	3,552	18.4	2.44	78	44
Unfurnished, other, rented	439	5.8	1,000	5.2	2.28	77	49
Furnished, rented	226	3.0	421	2.2	1.86	65	50
Rent free	97	1.3	210	1.1	2.16	87	45
Owns outright	1,913	25.3	3,887	20.0	2.03	87	55
Owns with mortgage	3,428	45.4	10,278	53.1	3.00	93	75
Analysis by age of main diary-keeper							
Age under 25	523	6.9	1,195	6.2	2.28	81	57
25-34	1,552	20.5	4,769	24.7	3.07	90	69
35-44	1,534	20.3	5,247	27.1	3.42	93	73
45-54	1,240	16.4	3,387	17.5	2.73	92	74
55-64	1,064	14.1	2,166	11.2	2.04	87	61
65-74	994	13.2	1,683	8.7	1.69	78	43
75 and over	634	8.4	874	4.5	1.38	69	22
Age unknown	15	0.2	27	0.1	1.80	100	47

(a) for definition of income groups see Table A2 of this Appendix and the Glossary
(b) see 'Adult' and 'Child' in the Glossary

Appendix B

Supplementary Tables

List of supplementary tables

		page
B1	Household consumption of individual foods: quarterly and annual national averages, 1992	68
B2	Average prices paid for household foods, 1990-1992	75
B3	Meals eaten outside the home, 1992	79
B4	Average number of mid-day meals per week per child aged 5-14 years, 1992	80
B5	Household food consumption of main food groups by income group, 1992	81
B6	Household food expenditure on main food groups by income group, 1992	83
B7	Household food expenditure on main food groups by household composition, 1992	85
B8	Household food consumption by household composition groups within income groups: selected food items, 1992	87
B9	Nutritional value of household food: national averages, 1990-1992	89
B10	Nutritional value of household food by region, 1992	90
B11	Nutritional value of household food by income group, 1992	91
B12	Nutritional value of household food by household composition, 1992	92
B13	Contributions made by selected foods to the nutritional value of household food: national averages, 1992	93

Table B1

Household consumption of individual foods: quarterly and annual national averages, 1992

ounces per person per week, except where otherwise stated

		Consumption Jan/March	April/June	July/Sept	Oct/Dec	Yearly average	Purchases Yearly average	Percentage of all households purchasing each type of food during Survey week
MILK AND CREAM:								
Liquid wholemilk, full price[a]	(pt)	1.84	1.69	1.66	1.57	1.69	1.66	53
Welfare milk	(pt)	0.03	0.03	0.03	0.04	0.03	...	...
School milk	(pt)	0.03	0.03	0.02	0.03	0.03	0.01	2
Low fat milks[a]	(pt)	1.57	1.70	1.67	1.85	1.70	1.70	55
Condensed milk	(eq pt)	0.04	0.05	0.06	0.05	0.05	0.05	6
Dried milk, branded	(eq pt)	0.04	0.03	0.05	0.07	0.05	0.04	1
Instant milk	(eq pt)	0.07	0.08	0.04	0.05	0.06	0.06	2
Other milks and dairy desserts[a]	(pt)	0.05	0.06	0.06	0.06	0.06	0.05	13
Yoghurt and fromage frais[a]	(pt)	0.20	0.21	0.22	0.22	0.21	0.21	37
Cream	(pt)	0.03	0.03	0.04	0.03	0.03	0.03	13
Total milk and cream	(pt or eq pt)	**3.89**	**3.90**	**3.84**	**3.98**	**3.91**	**3.82**	**95**
CHEESE:								
Natural[a]		3.61	3.76	3.51	3.73	3.66	3.65	56
Processed		0.34	0.33	0.38	0.36	0.35	0.35	12
Total cheese		**3.95**	**4.09**	**3.89**	**4.09**	**4.01**	**4.00**	**60**
MEAT AND MEAT PRODUCTS:								
Carcase meat:								
Beef and veal[a]		5.22	4.62	4.56	5.46	4.98	4.95	41
Mutton and lamb[a]		2.43	2.77	2.38	2.38	2.49	2.48	20
Pork[a]		2.58	2.48	2.52	2.54	2.53	2.51	23
Total carcase meat		**10.23**	**9.87**	**9.46**	**10.37**	**10.00**	**9.94**	**57**
Other meat and meat products:								
Liver[a]		0.35	0.24	0.25	0.27	0.28	0.28	6
Offals, other than liver		0.16	0.12	0.10	0.12	0.13	0.13	2
Bacon and ham, uncooked[a]		2.76	2.89	2.55	2.68	2.73	2.72	42
Bacon and ham, cooked, including canned		1.02	1.22	1.29	1.12	1.16	1.16	36
Cooked poultry, not purchased in cans		0.50	0.54	0.56	0.47	0.52	0.51	13
Corned meat		0.62	0.75	0.83	0.67	0.71	0.71	17
Other cooked meat, not purchased in cans		0.35	0.33	0.37	0.36	0.35	0.35	13
Other canned meat and canned meat products		1.33	1.23	1.49	1.31	1.33	1.33	16
Broiler chicken, uncooked, including frozen		5.26	5.12	4.86	4.97	5.06	5.05	29
Other poultry, uncooked, including frozen[a]		2.60	2.92	2.30	2.45	2.58	2.55	10
Rabbit and other meat		0.07	0.02	0.01	0.15	0.06	0.06	...
Sausages, uncooked, pork		1.26	1.28	1.26	1.31	1.28	1.27	18
Sausages, uncooked, beef		0.88	0.85	0.90	0.93	0.89	0.89	11
Meat pies and sausage rolls, ready-to-eat[a]		0.52	0.63	0.63	0.50	0.57	0.57	12
Frozen convenience meats or frozen convenience meat products[a]		2.63	2.55	2.56	2.31	2.51	2.51	24
Pate and delicatessan-type sausage[a]		0.32	0.29	0.29	0.31	0.30	0.30	11
Other meat products[a]		2.92	2.95	3.29	3.13	3.07	3.06	38
Total other meat and meat products		**23.56**	**23.92**	**23.54**	**23.06**	**23.52**	**23.45**	**86**
Total meat and meat products		**33.79**	**33.79**	**33.00**	**33.43**	**33.52**	**33.39**	**89**
FISH:								
White, filleted, fresh		0.62	0.61	0.59	0.62	0.61	0.61	10
White, unfilleted, fresh		0.12	0.05	0.05	0.03	0.06	0.05	...
White, uncooked, frozen		0.70	0.66	0.43	0.52	0.58	0.58	8
Herrings, filleted, fresh		0.02	0.03	0.01	0.02	0.02	0.02	...
Herrings, unfilleted, fresh		0.02	0.03	–	–	0.01	0.01	...
Fat, fresh, other than herrings		0.25	0.44	0.27	0.30	0.32	0.30	4
White, processed		0.14	0.14	0.18	0.17	0.15	0.15	3
Fat, processed, filleted		0.19	0.16	0.18	0.18	0.18	0.18	4
Fat, processed, unfilleted		0.01	...	–	...	...	...	...
Shellfish		0.16	0.24	0.26	0.21	0.22	0.22	4
Cooked fish		0.45	0.58	0.71	0.59	0.58	0.58	11
Canned salmon		0.24	0.35	0.40	0.24	0.30	0.30	8
Other canned or bottled fish		0.72	0.85	0.86	0.84	0.82	0.82	18
Fish products, not frozen		0.17	0.20	0.24	0.26	0.22	0.22	7
Frozen convenience fish products		0.85	0.93	0.95	0.96	0.92	0.92	13
Total fish		**4.65**	**5.28**	**5.12**	**4.95**	**4.99**	**4.97**	**58**

Table B1 *continued*

ounces per person per week, except where otherwise stated

		Consumption					Purchases	Percentage of all households purchasing each type of food during Survey week
		Jan/ March	April/ June	July/ Sept	Oct/ Dec	Yearly average	Yearly average	
EGGS	(no)	**2.23**	**2.16**	**1.96**	**1.97**	**2.08**	**2.03**	**49**
FATS:								
Butter[a]		1.40	1.47	1.40	1.49	1.44	1.43	23
Margarine[a]		2.96	2.77	2.65	2.78	2.79	2.79	29
Lard and compound cooking fat		0.63	0.53	0.52	0.60	0.57	0.57	9
Vegetable and salad oils	(fl oz)	1.78	1.70	1.40	2.02	1.73	1.73	10
Other fats[a]		2.19	2.00	2.19	2.08	2.11	2.11	27
Total fats		**8.97**	**8.46**	**8.17**	**8.97**	**8.66**	**8.65**	**66**
SUGAR AND PRESERVES:								
Sugar		5.75	5.22	5.74	5.35	5.51	5.51	30
Jams, jellies and fruit curds		0.72	0.67	0.64	0.78	0.70	0.68	11
Marmalade		0.49	0.57	0.60	0.57	0.56	0.55	8
Syrup, treacle		0.15	0.10	0.08	0.12	0.11	0.11	1
Honey		0.22	0.22	0.21	0.22	0.22	0.21	3
Total sugar and preserves		**7.32**	**6.78**	**7.27**	**7.04**	**7.10**	**7.07**	**42**
VEGETABLES:								
Old potatoes January-August		31.57	21.05	4.65	–	14.65	14.45	na
New potatoes January-August		1.48	9.65	16.82	–	6.70	6.32	na
Potatoes September-December		–	–	8.42	33.26	10.43	9.99	na
Total fresh potatoes		**33.05**	**30.70**	**29.89**	**33.26**	**31.78**	**30.74**	**58**
Fresh green vegetables:								
Cabbages, fresh		2.44	2.44	2.20	2.11	2.30	2.12	19
Brussels sprouts, fresh		1.46	0.07	0.25	1.58	0.86	0.81	10
Cauliflowers, fresh		2.56	3.14	3.37	2.99	3.00	2.91	26
Leafy salads, fresh		1.35	2.68	2.35	1.24	1.89	1.83	31
Peas, fresh		0.06	0.15	0.25	0.04	0.12	0.10	2
Beans, fresh		0.08	0.16	1.77	0.25	0.53	0.25	3
Other fresh green vegetables		0.07	0.18	0.09	0.05	0.10	0.07	1
Total fresh green vegetables		**8.03**	**8.83**	**10.30**	**8.26**	**8.81**	**8.09**	**58**
Other fresh vegetables:								
Carrots, fresh		4.65	3.76	3.55	4.33	4.09	3.91	38
Turnips and swedes, fresh		1.53	0.76	0.59	1.48	1.10	1.04	10
Other root vegetables, fresh		0.99	0.42	0.65	1.02	0.77	0.68	11
Onions, shallots, leeks, fresh		3.57	2.96	2.73	3.56	3.22	3.00	35
Cucumbers, fresh		0.85	1.69	1.57	0.91	1.24	1.18	23
Mushrooms, fresh		1.08	1.12	1.03	1.14	1.09	1.08	28
Tomatoes, fresh		2.58	4.04	4.26	2.53	3.32	3.04	43
Miscellaneous fresh vegetables		1.35	1.90	2.58	1.83	1.90	1.76	22
Total other fresh vegetables		**16.59**	**16.66**	**16.96**	**16.78**	**16.74**	**15.69**	**74**
Processed vegetables:								
Tomatoes, canned or bottled		1.82	1.68	1.55	1.84	1.73	1.73	19
Canned peas		1.43	1.40	1.44	1.44	1.43	1.43	17
Canned beans		4.72	4.41	3.83	3.95	4.24	4.24	36
Canned vegetables, other than pulses, potatoes or tomatoes		1.26	1.22	1.19	1.07	1.18	1.18	17
Dried pulses, other than air-dried		0.23	0.31	0.16	0.25	0.24	0.24	2
Air-dried vegetables		0.01	0.01	0.02	0.02	0.02	0.02	1
Vegetable juices	(fl oz)	0.35	0.24	0.22	0.31	0.28	0.28	5
Chips, excluding frozen		0.69	0.82	0.98	0.84	0.83	0.82	16
Instant potato		0.05	0.06	0.09	0.06	0.06	0.06	2
Canned potato		0.29	0.34	0.27	0.25	0.29	0.29	3
Potato products, not frozen[a]		1.38	1.46	1.58	1.49	1.47	1.47	37
Other vegetable products		1.07	1.50	1.37	1.07	1.25	1.25	24

Table B1 *continued*

ounces per person per week, except where otherwise stated

	Consumption					Purchases	Percentage of all households purchasing each type of food during Survey week
	Jan/March	April/June	July/Sept	Oct/Dec	Yearly average	Yearly average	
VEGETABLES *continued*							
Frozen peas	1.38	1.79	1.32	1.31	1.45	1.44	11
Frozen beans	0.44	0.41	0.35	0.38	0.39	0.37	3
Frozen chips and other frozen convenience potato products	3.30	3.25	3.18	3.29	3.26	3.26	18
All frozen vegetables and frozen vegetable products, not specified elsewhere	2.13	2.07	1.66	1.73	1.90	1.89	14
Total processed vegetables	**20.54**	**20.97**	**19.21**	**19.31**	**20.03**	**19.97**	**79**
Total vegetables	**78.21**	**77.15**	**76.36**	**77.61**	**77.36**	**74.49**	**93**
FRUIT:							
Fresh:							
Oranges	3.34	3.04	1.97	1.68	2.53	2.53	18
Other citrus fruit	2.83	1.61	1.06	2.84	2.11	2.11	20
Apples	6.37	6.44	6.26	7.24	6.59	6.32	46
Pears	1.27	1.32	1.11	1.41	1.28	1.27	14
Stone fruit	0.53	0.87	3.78	0.37	1.32	1.25	12
Grapes	0.96	0.88	1.00	1.23	1.02	1.01	13
Soft fruit, other than grapes	0.07	1.07	1.19	0.06	0.58	0.40	6
Bananas	4.36	5.46	5.42	5.06	5.06	5.06	44
Rhubarb	0.09	0.39	0.15	0.01	0.16	0.04	...
Other fresh fruit	0.77	1.01	1.77	1.13	1.15	1.15	9
Total fresh fruit	**20.61**	**22.09**	**23.71**	**21.02**	**21.80**	**21.15**	**71**
Other fruit and fruit products:							
Canned peaches, pears and pineapples	0.86	0.88	1.07	0.94	0.94	0.94	12
Other canned or bottled fruit	0.79	0.97	0.85	0.91	0.88	0.88	11
Dried fruit and dried fruit products	0.42	0.59	0.56	1.21	0.70	0.70	8
Frozen fruit and frozen fruit products	0.11	0.12	0.04	0.08	0.09	0.05	1
Nuts and nut products	0.39	0.44	0.49	0.92	0.56	0.56	12
Fruit juices (fl oz)	6.77	8.07	8.28	8.18	7.81	7.80	28
Total other fruit and fruit products	**9.35**	**11.09**	**11.32**	**12.27**	**10.99**	**10.93**	**48**
Total fruit	**29.95**	**33.18**	**35.03**	**33.29**	**32.79**	**32.08**	**79**
CEREALS:							
White bread, standard loaves, unsliced	2.44	2.49	2.14	2.06	2.29	2.29	17
White bread, standard loaves, sliced	11.49	10.49	10.88	9.78	10.66	10.65	43
White sliced premium loaves	0.93	1.19	1.56	1.35	1.25	1.25	7
White sliced softgrain loaves	0.99	1.06	0.80	0.92	0.95	0.95	5
Brown bread	3.48	3.46	3.10	3.08	3.29	3.29	23
Wholemeal bread	3.67	3.85	3.95	3.98	3.86	3.86	24
Other bread[a]	3.77	4.49	4.66	4.44	4.33	4.33	47
Total bread	**26.77**	**27.04**	**27.11**	**25.61**	**26.62**	**26.61**	**90**
Flour	3.20	2.85	2.67	2.61	2.84	2.84	11
Buns, scones and teacakes	1.35	1.45	1.27	1.53	1.40	1.40	24
Cakes and pastries	2.38	2.73	2.63	2.99	2.68	2.68	39
Crispbread	0.22	0.18	0.16	0.19	0.19	0.19	6
Biscuits, other than chocolate biscuits	3.57	3.46	3.39	3.49	3.48	3.48	47
Chocolate biscuits	1.40	1.36	1.55	1.92	1.56	1.55	31
Oatmeal and oat products	0.67	0.48	0.43	0.52	0.53	0.53	6
Breakfast cereals[a]	4.48	4.59	4.96	4.62	4.66	4.66	43
Canned milk puddings	0.92	0.93	0.83	0.93	0.90	0.90	10
Other puddings	0.16	0.08	0.11	0.31	0.16	0.16	3
Rice	1.14	1.42	1.59	1.21	1.33	1.33	9
Cereal-based invalid foods (including 'slimming' foods)	0.01	0.02	0.05	0.01	0.02	0.02	...
Infant cereal foods	0.06	0.05	0.06	0.06	0.06	0.06	1
Frozen convenience cereal foods[a]	1.29	1.32	1.43	1.36	1.35	1.35	16
Cereal convenience foods, including canned, not specified elsewhere[a]	2.87	2.79	2.75	2.92	2.84	2.84	41
Other cereal foods	0.89	0.99	0.96	0.95	0.95	0.95	11
Total cereals	**51.38**	**51.76**	**51.95**	**51.20**	**51.57**	**51.55**	**95**

Table B1 *continued*

ounces per person per week, except where otherwise stated

		Consumption					Purchases	Percentage of all households purchasing each type of food during Survey week
		Jan/ March	April/ June	July/ Sept	Oct/ Dec	Yearly average	Yearly average	
BEVERAGES:								
Tea		1.45	1.39	1.29	1.33	1.37	1.37	32
Coffee, bean and ground		0.13	0.16	0.12	0.16	0.14	0.14	3
Coffee, instant		0.48	0.50	0.46	0.50	0.49	0.49	23
Coffee, essences	(fl oz)	–	0.01	–	0.01	...	...	...
Cocoa and drinking chocolate		0.16	0.12	0.10	0.10	0.12	0.12	3
Branded food drinks		0.26	0.22	0.20	0.23	0.23	0.23	5
Total beverages		**2.47**	**2.41**	**2.17**	**2.33**	**2.35**	**2.35**	**50**
MISCELLANEOUS:								
Mineral water[b]	(fl oz)	2.77	4.93	3.49	2.55	3.43	3.42	7
Baby foods, canned or bottled		0.17	0.29	0.20	0.23	0.22	0.22	2
Soups, canned		2.98	2.04	1.95	2.82	2.47	2.46	21
Soups, dehydrated and powdered		0.14	0.09	0.08	0.13	0.11	0.11	6
Accelerated freeze-dried foods (excluding coffee)		0.01	–	–	–	–	–	...
Spreads and dressings[a]		0.52	0.83	0.78	0.60	0.68	0.68	12
Pickles and sauces		2.36	2.66	2.51	2.68	2.55	2.54	32
Meat and yeats extracts		0.17	0.15	0.11	0.15	0.15	0.15	9
Table jellies, squares and crystals		0.16	0.22	0.16	0.17	0.17	0.17	6
Ice-cream, mousse	(fl oz)	1.85	3.12	3.13	2.24	2.57	2.56	11
Ice-cream products and other frozen dairy foods[a]		0.58	0.98	0.94	0.46	0.74	0.74	8
Salt		0.41	0.45	0.46	0.36	0.42	0.42	4
Novel protein foods		0.09	0.04	0.05	0.06	0.06	0.06	1
SOFT DRINKS:								
Soft drinks, concentrated	(fl oz)	4.30	5.01	4.42	4.08	4.45	4.45	21
Soft drinks, unconcentrated	(fl oz)	12.33	14.74	15.33	14.05	14.07	14.04	37
Low calorie soft drinks, concentrated	(fl oz)	0.29	0.51	0.98	0.70	0.61	0.61	3
Low calorie soft drinks, unconcentrated	(fl oz)	4.90	7.04	7.23	6.00	6.26	6.26	15
Total soft drinks	(fl oz)	**21.82**	**27.29**	**27.96**	**24.83**	**25.39**	**25.36**	**54**
ALCOHOLIC DRINKS:								
Low alcohol beers, lagers and ciders	(cl)	0.43	0.37	0.17	0.18	0.29	0.29	...
Beers	(cl)	5.65	6.88	7.23	5.76	6.35	6.35	6
Lagers and continental beers	(cl)	7.94	10.50	10.99	11.26	10.14	10.04	8
Ciders and perry	(cl)	1.20	2.59	2.31	1.93	1.99	1.99	3
Wine	(cl)	6.78	8.65	8.75	10.81	8.74	8.69	14
LA wine, wine and spirits with additions	(cl)	0.30	0.35	0.30	0.27	0.31	0.31	1
Fortified wines	(cl)	1.32	1.05	1.21	1.80	1.35	1.35	3
Spirits	(cl)	1.48	1.41	1.42	1.67	1.50	1.50	4
Liqueurs	(cl)	0.08	0.15	0.04	0.21	0.12	0.12	...
Total alcoholic drinks	(cl)	**25.15**	**31.95**	**32.41**	**33.89**	**30.78**	**30.64**	**27**
CONFECTIONERY:								
Solid chocolate	(g)	8.45	11.27	10.79	14.59	11.27	11.27	14
Chocolate coated filled bar/sweets	(g)	21.80	22.93	21.11	28.82	23.72	23.71	21
Chewing gum	(g)	0.47	0.38	0.31	0.19	0.34	0.34	2
Mints and boiled sweets[a]	(g)	14.42	13.44	14.40	12.95	13.79	13.79	15
Fudge, toffees, caramels	(g)	1.93	1.83	1.60	1.82	1.80	1.80	2
Total confectionery	(g)	**47.07**	**49.85**	**48.22**	**58.35**	**50.92**	**50.90**	**37**

(a) these foods are given in greater detail in this table under 'Supplementary classifications'
(b) the second quarter figure and yearly average are inflated by a large purchase observed in a single household this year

Table B1 continued

ounces per person per week, except where otherwise stated

Supplementary classifications[c]		Consumption					Purchases	Percentage of all households purchasing each type of food during Survey week
		Jan/ March	April/ June	July/ Sept	Oct/ Dec	Yearly average	Yearly average	
MILK AND CREAM:								
Liquid wholemilk, full price:								
UHT	(pt)	0.04	0.01	0.02	0.01	0.02	0.02	1
sterilised	(pt)	0.09	0.08	0.12	0.07	0.09	0.09	4
other	(pt)	1.71	1.60	1.51	1.49	1.58	1.55	50
Total liquid wholemilk, full price	(pt)	**1.84**	**1.69**	**1.66**	**1.57**	**1.69**	**1.66**	53
Low fat milks:								
fully skimmed	(pt)	0.34	0.37	0.41	0.38	0.37	0.37	16
semi and other skimmed	(pt)	1.23	1.33	1.26	1.47	1.32	1.32	44
Total skimmed milks	(pt)	**1.57**	**1.70**	**1.67**	**1.85**	**1.70**	**1.70**	55
Other milks and dairy desserts:								
dairy desserts	(pt)	0.02	0.03	0.03	0.03	0.03	0.03	7
other milks	(pt)	0.03	0.03	0.03	0.03	0.03	0.02	6
Total other milks	(pt)	**0.05**	**0.06**	**0.06**	**0.06**	**0.06**	**0.05**	13
Yoghurt and fromage frais:								
yoghurt	(pt)	0.18	0.19	0.20	0.20	0.19	0.19	34
fromage frais	(pt)	0.02	0.02	0.02	0.02	0.02	0.02	6
Total yoghurt and fromage frais	(pt)	**0.20**	**0.21**	**0.22**	**0.22**	**0.21**	**0.21**	37
CHEESE:								
Natural hard:								
Cheddar and Cheddar type		2.40	2.45	2.26	2.38	2.38	2.38	41
other UK varieties or foreign equivalents		0.49	0.58	0.56	0.62	0.56	0.56	13
Edam and other continental		0.28	0.27	0.22	0.28	0.26	0.26	7
Cottage		0.28	0.29	0.31	0.23	0.27	0.27	7
Other natural soft		0.17	0.16	0.17	0.22	0.18	0.18	6
Total natural cheese		**3.61**	**3.76**	**3.51**	**3.73**	**3.66**	**3.65**	56
CARCASE MEAT:								
Beef: joints (including sides) on the bone		0.21	0.17	0.30	0.10	0.19	0.19	1
joints, boned		1.19	0.97	0.99	1.38	1.14	1.12	8
steak, less expensive varieties		1.18	0.77	0.95	1.24	1.04	1.04	13
steak, more expensive varieties		0.64	0.81	0.60	0.67	0.68	0.68	10
minced		1.97	1.86	1.71	2.05	1.90	1.90	23
other, and veal		0.03	0.04	...	0.02	0.02	0.02	...
Total beef and veal		**5.22**	**4.62**	**4.56**	**5.46**	**4.98**	**4.95**	41
Mutton		0.02	0.01	0.01	0.02	0.02	0.02	...
Lamb: joints (including sides)		1.33	1.65	1.24	1.31	1.39	1.38	8
chops (including cutlets and fillets)		0.84	0.87	0.98	0.88	0.89	0.89	11
all other		0.24	0.23	0.15	0.17	0.20	0.20	2
Total mutton and lamb		**2.43**	**2.77**	**2.38**	**2.38**	**2.49**	**2.48**	20
Pork: joints (including sides)		0.83	0.64	0.67	0.87	0.75	0.75	4
chops		1.17	1.09	1.20	1.04	1.12	1.12	13
fillets and steaks		0.27	0.41	0.31	0.28	0.32	0.32	4
all other		0.31	0.33	0.34	0.36	0.33	0.33	4
Total pork		**2.58**	**2.48**	**2.52**	**2.54**	**2.53**	**2.51**	23
OTHER MEAT AND MEAT PRODUCTS:								
Liver: ox		0.03	0.01	0.01	0.02	0.02	0.02	...
lambs		0.20	0.14	0.17	0.16	0.17	0.17	4
pigs		0.10	0.06	0.05	0.08	0.08	0.08	2
other		0.02	0.03	0.01	0.01	0.02	0.02	...
Total liver		**0.35**	**0.25**	**0.25**	**0.27**	**0.28**	**0.28**	6

Table B1 *continued*

ounces per person per week, except where otherwise stated

	Consumption					Purchases	Percentage of all households purchasing each type of food during Survey week
Supplementary classifications[c]	Jan/ March	April/ June	July/ Sept	Oct/ Dec	Yearly average	Yearly average	
OTHER MEAT AND MEAT PRODUCTS *continued*							
Bacon and ham, uncooked:							
joints (including sides and steaks cut from joint)	0.63	0.75	0.45	0.73	0.65	0.65	7
rashers, vacuum-packed	0.74	0.77	0.84	0.86	0.80	0.80	16
rashers, not vacuum-packed	1.38	1.38	1.26	1.09	1.28	1.28	23
Total bacon and ham, uncooked	**2.76**	**2.89**	**2.55**	**2.68**	**2.73**	**2.72**	**42**
Poultry, uncooked, including frozen:							
chicken, other than broilers	1.93	1.84	1.52	1.36	1.67	1.66	5
turkey	0.59	1.01	0.68	0.92	0.80	0.79	5
all other	0.08	0.06	0.11	0.18	0.11	0.10	...
Total poultry, uncooked, including frozen	**2.60**	**2.92**	**2.30**	**2.46**	**2.58**	**2.55**	**10**
Meat pies and sausage rolls, ready-to-eat:							
meat pies	0.35	0.47	0.44	0.35	0.40	0.40	9
sausage rolls	0.16	0.17	0.19	0.15	0.17	0.17	4
Total meat pies and sausage rolls, ready-to-eat	**0.52**	**0.63**	**0.63**	**0.50**	**0.57**	**0.57**	**12**
Frozen convenience meats or frozen convenience meat products:							
burgers	0.72	0.76	0.78	0.57	0.71	0.71	9
meat pies, pasties, puddings	0.64	0.45	0.66	0.52	0.57	0.57	6
other	1.28	1.33	1.12	1.21	1.24	1.24	15
Total frozen convenience meats or frozen convenience meat products	**2.63**	**2.55**	**2.56**	**2.31**	**2.51**	**2.51**	**24**
Pate and delicatessan-type sausages:							
pate	0.11	0.13	0.11	0.11	0.11	0.11	5
delicatessen-type sausages	0.21	0.16	0.18	0.20	0.19	0.19	7
Total pate and delicatessen-type sausages	**0.32**	**0.29**	**0.29**	**0.31**	**0.30**	**0.30**	**11**
Other meat products:							
meat pastes and spreads	0.07	0.08	0.07	0.07	0.07	0.07	5
meat pies, pasties and puddings	1.24	0.98	1.18	1.17	1.14	1.14	18
ready meals	0.88	1.05	1.04	1.03	1.00	0.99	12
other meat products, not specified elsewhere	0.73	0.84	1.01	0.85	0.85	0.85	15
Total other meat products	**2.92**	**2.95**	**3.29**	**3.13**	**3.07**	**3.06**	**38**
FATS:							
Butter: New Zealand	0.51	0.45	0.40	0.41	0.44	0.44	7
Danish	0.27	0.24	0.29	0.34	0.29	0.29	5
UK	0.16	0.19	0.16	0.19	0.17	0.17	3
other	0.46	0.58	0.55	0.55	0.53	0.53	9
Total butter	**1.40**	**1.47**	**1.40**	**1.49**	**1.44**	**1.43**	**23**
Margarine: soft	2.75	2.59	2.53	2.58	2.62	2.62	27
other	0.21	0.17	0.12	0.20	0.18	0.18	3
Total margarine	**2.96**	**2.77**	**2.65**	**2.78**	**2.79**	**2.79**	**29**
Other fats:							
reduced fat spreads	0.90	0.91	0.93	0.83	0.89	0.89	11
low fat spreads	0.97	0.87	0.88	0.92	0.91	0.91	13
suet and dripping	0.06	0.03	0.06	0.08	0.06	0.06	1
other fats	0.25	0.19	0.32	0.25	0.25	0.25	5
Total other fats	**2.19**	**2.00**	**2.19**	**2.08**	**2.11**	**2.11**	**27**

Table B1 *continued*

ounces per person per week, except where otherwise stated

Supplementary classifications(c)		Consumption					Purchases	Percentage of all households purchasing each type of food during Survey week
		Jan/ March	April/ June	July/ Sept	Oct/ Dec	Yearly average	Yearly average	
VEGETABLES:								
Potato products, not frozen:								
crisps and potato snacks		1.29	1.33	1.46	1.35	1.35	1.35	36
other potato products, not frozen		0.10	0.13	0.11	0.14	0.12	0.12	3
Potato products, not frozen		**1.38**	**1.46**	**1.58**	**1.49**	**1.47**	**1.47**	37
CEREALS:								
Other bread:								
rolls (excluding starch reduced rolls)		2.14	2.60	2.54	2.46	2.43	2.43	30
malt bread and fruit bread		0.21	0.18	0.23	0.17	0.20	0.20	4
Vienna bread and French bread		0.45	0.72	0.70	0.80	0.66	0.66	10
starch reduced bread and rolls		0.34	0.35	0.42	0.25	0.34	0.34	4
sandwiches		0.10	0.10	0.17	0.14	0.13	0.13	3
other		0.53	0.56	0.60	0.63	0.58	0.58	9
Total other bread		**3.77**	**4.49**	**4.66**	**4.44**	**4.33**	**4.33**	47
Breakfast cereals:								
muesli		0.64	0.63	0.78	0.69	0.68	0.68	6
other high fibre breakfast cereals		1.88	1.88	1.86	1.97	1.90	1.90	22
sweetened breakfast cereals		0.82	0.87	0.90	0.77	0.83	0.83	12
other breakfast cereal		1.15	1.22	1.42	1.20	1.24	1.24	16
Total breakfast cereals		**4.48**	**4.60**	**4.96**	**4.62**	**4.66**	**4.66**	43
Frozen convenience cereal foods:								
cakes and pastries		0.49	0.52	0.59	0.62	0.56	0.56	8
other		0.80	0.80	0.84	0.74	0.79	0.79	10
Total frozen convenience cereal foods		**1.29**	**1.32**	**1.43**	**1.36**	**1.35**	**1.35**	16
Cereal convenience foods, including canned, not specified elsewhere:								
canned pasta		1.33	1.18	1.22	1.22	1.24	1.24	15
pizza		0.49	0.54	0.54	0.53	0.52	0.52	7
cake, pudding and dessert mixes		0.37	0.32	0.34	0.35	0.34	0.34	10
other		0.68	0.76	0.66	0.82	0.73	0.73	23
Total cereal convenience foods, including canned, not specified elsewhere		**2.87**	**2.79**	**2.75**	**2.92**	**2.84**	**2.84**	41
MISCELLANEOUS:								
Spreads and dressings:								
salad dressings		0.45	0.75	0.70	0.49	0.59	0.59	11
other spreads and dressings		0.06	0.08	0.07	0.11	0.08	0.08	2
Total spreads and dressings		**0.52**	**0.83**	**0.78**	**0.60**	**0.68**	**0.68**	12
Ice-cream products and other frozen dairy foods:								
ice-cream products		0.40	0.76	0.69	0.35	0.55	0.54	7
other frozen dairy foods		0.18	0.22	0.25	0.11	0.19	0.19	2
Total ice-cream products and other frozen dairy foods		**0.58**	**0.98**	**0.94**	**0.46**	**0.74**	**0.74**	8
CONFECTIONERY:								
Mints and boiled sweets:								
hard pressed mints	(g)	1.70	0.95	1.65	0.87	1.29	1.29	3
boiled sweets	(g)	12.73	12.49	12.75	12.08	12.51	12.50	13
Total mints and boiled sweets	(g)	**14.42**	**13.44**	**14.40**	**12.95**	**13.79**	**13.79**	15

(c) supplementary data for certain foods in greater detail than shown elsewhere in the table; the totals for each main food are repeated for ease of reference

Table B2
Average prices paid[a] for household foods, 1990 - 1992

pence per lb [b]

	Average prices paid		
	1990	1991	1992
MILK AND CREAM:			
Liquid wholemilk, full price	29.87	31.33	31.99
Low fat milks	29.02	30.39	30.98
Dried milk, branded	39.86	43.25	44.51
Instant milk	19.56	21.73	21.30
Other milk	71.57	70.87	71.42
Yoghurt and fromage frais	96.59	102.81	108.57
Cream	162.11	142.81	135.10
CHEESE:			
Natural	163.36	166.55	181.75
Processed	195.87	201.05	205.84
MEAT AND MEAT PRODUCTS:			
Carcase meat			
Beef and veal	196.15	203.39	201.68
Mutton and lamb	158.22	155.37	171.33
Pork	157.52	153.95	157.69
Other meat and meat products			
Liver	100.41	97.37	102.36
Offals, other than liver	111.15	109.92	104.26
Bacon and ham, uncooked	179.80	176.53	188.38
Bacon and ham, cooked, including canned	253.88	258.00	268.98
Cooked poultry, not purchased in cans	233.55	240.53	259.86
Corned meat	145.40	144.70	125.20
Other cooked meat, not purchased in cans	241.39	258.02	273.84
Other canned meat and canned meat products	87.84	94.72	90.96
Broiler chicken, uncooked, including frozen	113.23	112.33	110.13
Other poultry, uncooked, including frozen	95.97	101.95	95.73
Rabbit and other meat	138.34	133.99	119.06
Sausages, uncooked, pork	108.28	112.67	113.24
Sausages, uncooked, beef	93.06	92.89	92.97
Meat pies and sausage rolls, ready-to-eat	152.11	140.90	147.69
Frozen convenience meats or frozen convenience meat products	155.65	156.43	157.06
Pate and delicatessen type sausages	212.49	200.67	214.07
Other meat products	230.24	221.49	241.12
FISH:			
White, filleted, fresh	220.12	252.35	258.44
White, unfilleted, fresh	163.06	164.31	166.39
White, uncooked, frozen	188.02	215.28	211.99
Herrings, filleted, fresh	126.98	117.07	124.67
Herrings, unfilleted, fresh	98.26	109.62	92.57
Fat, fresh, other than herrings	240.29	230.58	243.41
White, processed	223.30	267.33	270.15
Fat, processed, filleted	211.08	212.60	251.30
Fat, processed, unfilleted	135.32	208.07	381.51
Shellfish	371.44	367.91	360.48
Cooked fish	328.40	330.23	300.84
Canned salmon	288.63	230.79	202.98
Other canned or bottled fish	124.84	128.01	122.89
Fish products, not frozen	276.05	271.37	306.71
Frozen convenience fish products	166.19	175.12	171.69
EGGS	9.21	9.22	9.25
FATS:			
Butter	109.79	108.84	111.93
Margarine	49.96	55.02	57.56
Low fat and dairy spreads	80.69	79.42	81.07
Vegetable and salad oils	49.11	51.49	56.29
Other fats	65.92	67.39	60.55

Table B2 *continued*

pence per lb(b)

		Average prices paid		
		1990	1991	1992
SUGAR AND PRESERVES:				
Sugar		29.40	31.61	30.82
Jams, jellies and fruit curds		65.80	74.38	78.76
Marmalade		63.20	70.10	73.48
Syrup, treacle		59.61	57.70	58.60
Honey		92.92	99.46	102.70
VEGETABLES:				
Old potatoes	January-August	13.33	12.14	12.62
New potatoes	January-August	16.44	20.13	17.32
Potatoes	September-December	10.04	12.01	11.02
Fresh:				
Cabbages		27.33	29.42	27.82
Brussels sprouts		31.70	37.16	31.89
Cauliflowers		33.19	42.97	38.74
Leafy salads		70.14	70.47	70.68
Peas		88.94	105.20	120.94
Beans		77.23	79.84	90.04
Other green vegetables		85.29	82.80	83.44
Carrots		25.70	27.27	22.88
Turnips and swedes		23.90	24.99	22.75
Other root vegetables		45.97	51.80	48.54
Onions, shallots, leeks		35.54	37.97	36.33
Cucumbers		62.11	57.25	58.04
Mushrooms		117.93	119.89	124.25
Tomatoes		63.56	65.36	59.54
Miscellaneous fresh vegetables		79.16	82.66	82.60
Processed:				
Tomatoes, canned or bottled		33.06	31.46	28.42
Canned peas		34.12	37.71	37.07
Canned beans		30.72	31.71	30.25
Canned vegetables, other than pulses, potatoes or tomatoes		52.15	53.00	55.00
Dried pulses, other than air-dried		62.76	47.49	50.66
Air-dried vegetables		250.97	145.15	159.53
Vegetables juices		81.09	87.72	84.34
Chips, excuding frozen		156.36	157.70	139.95
Instant potato		114.07	140.81	135.76
Canned potato		38.87	40.95	39.41
Potato products, not frozen		205.23	221.40	230.48
Other vegetable products		151.90	154.81	165.04
Frozen peas		48.37	48.55	49.37
Frozen beans		49.07	57.26	60.04
Frozen chips and other frozen convenience potato products		43.67	44.92	43.84
All frozen vegetable and frozen vegetable products, not specified elsewhere		64.74	72.71	74.63
FRUIT:				
Fresh:				
Oranges		38.00	39.54	39.04
Other citrus fruit		47.69	49.78	50.49
Apples		44.99	51.67	49.52
Pears		49.93	52.53	49.67
Stone fruit		73.83	78.35	68.78
Grapes		94.65	94.45	90.67
Soft fruit, other than grapes		108.10	116.84	120.68
Bananas		49.69	51.43	46.38
Rhubarb		33.67	43.59	44.15
Other fresh fruit		59.29	53.29	55.84

Table B2 *continued*

pence per lb[b]

	Average prices paid		
	1990	1991	1992
FRUIT *continued*			
Canned peaches, pears and pineapples	44.10	46.62	47.95
Other canned or bottled fruit	55.45	62.42	64.21
Dried fruit and dried fruit products	84.35	94.28	100.74
Frozen fruits and frozen fruit products	126.17	117.01	132.69
Nuts and nut products	152.94	166.66	160.51
Fruit juices	48.56	48.13	46.84
CEREALS:			
White bread, standard loaves, unsliced	35.71	42.15	44.48
White bread, standard loaves, sliced	28.56	29.44	27.20
White bread, sliced, premium	na	na	35.11
White bread, sliced, softgrain	na	na	33.32
Brown bread	38.51	40.40	40.46
Wholemeal bread	37.47	40.36	38.92
Other bread	60.80	66.69	83.21
Flour	17.75	19.92	21.04
Buns, scones and teacakes	87.86	88.69	89.46
Cakes and pastries	145.82	130.57	139.21
Crispbread	96.36	109.16	118.45
Biscuits, other than chocoloate	77.51	82.16	84.94
Chocolate biscuits	144.28	143.13	146.03
Oatmeal and oat products	68.21	65.80	50.28
Breakfast cereals	94.77	101.36	108.65
Canned milk puddings	41.93	46.02	45.85
Other puddings	144.57	143.75	143.80
Rice	58.57	60.48	66.16
Cereal-based invalid foods (including 'slimming' foods)	251.55	337.03	475.69
Infant cereal foods	278.14	315.26	350.01
Frozen convenience cereal foods	140.08	148.56	154.31
Cereal convenience foods, including canned, not specified elsewhere	125.59	134.69	146.02
Other cereal foods	62.70	64.47	64.31
BEVERAGES:			
Tea	211.63	233.21	227.27
Coffee, bean and ground	261.25	256.39	263.63
Coffee, instant	582.06	551.35	557.99
Coffee, essences	231.34	239.82	222.82
Cocoa and drinking chocolate	167.23	150.42	152.16
Branded food drinks	161.69	197.63	220.42
MISCELLANEOUS:			
Mineral water	21.44	23.56	24.73
Baby foods, canned or bottled	101.46	118.40	113.66
Soups, canned	38.34	43.52	46.03
Soups, dehydrated and powdered	299.32	281.69	335.79
Accelerated freeze-dried foods (excluding coffee)	–	–	32.91
Spreads and dressings	91.86	99.71	103.53
Pickles and sauces	76.48	85.22	92.28
Meat and yeast extracts	307.10	331.31	353.28
Table jellies, squares and crystals	83.63	98.53	102.58
Ice-cream, mousse	56.87	80.30	63.59
Ice-cream products and other frozen dairy foods	na	na	135.13
Salt	22.93	24.20	27.36
Novel protein foods	212.03	177.62	194.59

Table B2 *continued*

pence per lb[b]

	Average prices paid		
	1990	1991	1992
SOFT DRINKS:			
Soft drinks, concentrated	44.08	46.97	53.00
Soft drinks, unconcentrated	27.43	28.92	29.85
Low calorie soft drinks, concentrated	na	na	47.22
Low calorie soft drinks, unconcentrated[c]	26.90	28.86	28.94
ALCOHOLIC DRINKS:			
Low alcohol beers, lagers and ciders	na	na	1.18
Beers	na	na	1.56
Lagers and continental beers	na	na	1.45
Ciders and perry	na	na	1.48
Wine	na	na	3.86
Low alcohol wine, wines and spirits with additions	na	na	2.63
Fortified wines	na	na	4.91
Spirits	na	na	13.78
Liqueurs	na	na	14.06
CONFECTIONERY:			
Solid Chocolate	na	na	49.62
Chocolate coated filled bar/sweets	na	na	48.92
Chewing gum	na	na	76.44
Mints and boiled sweets	na	na	39.56
Fudge, toffees, caramels	na	na	39.15

(a) it should be noted that since the results for household consumption presented in this Report include both purchases and 'free food', average prices paid cannot in general be derived by dividing the expenditure on a particular food by average consumption

(b) pence per lb, except for the following; per pint of milk, yoghurt, cream, vegetable and salad oils, vegetable juices, fruit juices, coffee essence, ice-cream, soft drinks; per equivalent pint of condensed, dried and instant milk; per 100 grams of confectionery; per centilitre of alcoholic drink; per egg

(c) includes some concentrated low calorie soft drinks before 1992

Table B3
Meals eaten outside the home, 1992

per person per week

	Meals not from the household supply		Net balance[a]	
	Mid-day meals	All meals out[b]	Persons	Visitors
All households	1.72	2.78	0.87	0.05
Analysis by region				
Scotland	1.90	3.19	0.85	0.06
Wales	1.66	2.70	0.87	0.04
England	1.71	2.75	0.87	0.05
North	1.83	2.84	0.87	0.04
Yorkshire and Humberside	1.77	2.84	0.87	0.05
North West	1.87	2.93	0.86	0.05
East Midlands	1.60	2.59	0.88	0.04
West Midlands	1.55	2.51	0.88	0.05
South West	1.46	2.39	0.88	0.04
South East/East Anglia	1.75	2.84	0.86	0.05
Analysis by income group				
A1	2.59	4.43	0.79	0.07
A2	2.31	3.77	0.82	0.05
B	1.95	3.12	0.85	0.04
C	1.78	2.87	0.86	0.04
D	1.55	2.41	0.89	0.05
E1	1.17	2.05	0.90	0.06
E2	1.48	2.31	0.89	0.06
OAP (households containing one adult)	1.06	1.84	0.91	0.06
OAP (households containing one male and one female)	0.51	0.93	0.96	0.04
OAP ('other' households)	0.81	1.57	0.93	0.08
OAP (all)	0.75	1.33	0.94	0.05
Analysis by household composition				
Number of adults / Number of children				
1 / 0	1.66	3.01	0.85	0.08
1 / 1 or more	2.63	3.78	0.82	0.06
2 / 0	1.44	2.50	0.88	0.06
2 / 1	1.87	2.99	0.86	0.05
2 / 2	1.81	2.76	0.87	0.03
2 / 3	1.65	2.39	0.89	0.03
2 / 4 or more	1.80	2.38	0.89	0.02
3 / 0	1.62	2.77	0.87	0.05
3 or more / 1 or 2	1.86	2.96	0.86	0.03
3 or more / 3 or more	1.86	2.90	0.86	0.02
4 or more / 0	1.89	3.13	0.85	0.04
Analysis by age of main diary-keeper				
Under 25 years	2.09	3.61	0.82	0.05
25-34 years	2.02	3.26	0.85	0.04
35-44 years	1.95	2.99	0.86	0.04
45-54 years	1.81	2.94	0.86	0.06
55-64 years	1.14	1.93	0.90	0.06
65-74 years	0.94	1.75	0.91	0.05
75 and over	0.82	1.35	0.93	0.03
Analysis by housing tenure				
Unfurnished: council	1.63	2.51	0.88	0.04
other, rented	1.69	2.63	0.88	0.05
Furnished, rented	2.52	4.25	0.79	0.06
Rent free	1.68	2.85	0.86	0.05
Owned outright	1.26	2.16	0.89	0.06
Owned with mortgage	1.90	3.07	0.85	0.04
Analysis by ownership of deep-freezer				
Households owning a deep-freezer	1.73	2.79	0.87	0.05
Households not owning a deep-freezer	1.62	2.66	0.87	0.04

(a) see Glossary

(b) based on a pattern of three meals a day; this differs from the results of the survey produced up to 1990 because less important fourth meals eaten out (i.e. afternoon or evening) are no longer included.

Table B4

Average number of mid-day meals per week per child aged 5-14 years, 1992

	Meals not from the household supply		Meals from the household supply	
	School meals	Other meals out	Packed meals	Other
All households	1.74	0.25	1.74	3.27
Analysis by region				
Scotland	1.47	0.42	1.14	3.97
Wales	1.85	0.30	1.36	3.49
England	1.76	0.23	1.82	3.19
North	2.17	0.33	1.06	3.44
Yorkshire and Humberside	1.97	0.22	1.77	3.04
North West	2.32	0.24	1.21	3.23
East Midlands	1.89	0.22	1.79	3.10
West Midlands	1.69	0.27	1.75	3.29
South West	1.33	0.14	2.33	3.20
South East/East Anglia	1.49	0.23	2.17	3.11
Analysis by income group				
A1	1.98	0.47	1.48	3.07
A2	1.30	0.24	2.13	3.33
B	1.37	0.30	2.08	3.25
C	1.60	0.23	1.87	3.30
D	2.21	0.17	1.30	3.32
E1	2.38	0.15	0.77	3.70
E2	3.14	0.19	0.55	3.12
OAP (all)	(a)	(a)	(a)	(a)
Analysis by household composition				
Number of adults / Number of children				
1 / 1 or more	2.97	0.21	0.90	2.92
2 / 1	1.68	0.32	1.71	3.29
2 / 2	1.52	0.28	1.96	3.24
2 / 3	1.50	0.22	2.00	3.28
2 / 4 or more	1.95	0.19	1.56	3.30
3 or more / 1 or 2	1.40	0.31	1.49	3.80
3 or more / 3 or more	2.10	0.14	1.34	3.42
Analysis by age of main diary-keeper				
Under 25 years	2.83	0.24	0.86	3.07
25-34 years	1.84	0.22	1.62	3.32
35-44 years	1.68	0.28	1.88	3.16
45-54 years	1.62	0.24	1.62	3.52
55-64 years	(a)	(a)	(a)	(a)
65-74 years	(a)	(a)	(a)	(a)
75 and over	(a)	(a)	(a)	(a)
Analysis by housing tenure				
Unfurnished: council	2.39	0.24	1.02	3.35
other, rented	2.45	0.29	1.30	2.96
Furnished, rented	2.21	0.31	1.45	3.03
Rent free	1.59	0.23	1.95	3.23
Owned outright	1.94	0.29	1.54	3.23
Owned with mortgage	1.45	0.25	2.03	3.27
Analysis by ownership of deep-freezer				
Households owning a deep-freezer	1.73	0.25	1.77	3.25
Households not owning a deep-freezer	1.98	0.30	1.01	3.71

(a) estimates are not shown because these household groups contain samples of fewer than 20 children aged 5-14 years

Table B5
Household food consumption of main food groups by income group, 1992

ounces per person per week, except where otherwise stated

		| Income groups								
		| Gross weekly income of head of household								
		| Households with one or more earners						Households without an earner		OAP
		£730 and over	£520 and under £730	£520 and over	£280 and under £520	£140 and under £280	Less than £140	£140 or more	Less than £140	
		A1	A2	All A	B	C	D	E1	E2	
MILK AND CREAM:										
Liquid wholemilk, full price	(pt)	1.13	1.25	1.21	1.38	1.74	1.89	1.81	2.04	2.45
Welfare and school milk	(pt)	0.01	0.02	0.02	0.03	0.03	0.10	0.04	0.31	–
Low fat milks	(pt)	2.33	2.01	2.11	1.86	1.55	1.51	2.11	1.40	1.56
Dried and other milk	(pt or eq pt)	0.51	0.46	0.47	0.44	0.38	0.37	0.57	0.48	0.46
Cream	(pt)	0.05	0.05	0.05	0.03	0.03	0.02	0.06	0.02	0.04
Total milk and cream	(pt or eq pt)	**4.04**	**3.79**	**3.87**	**3.73**	**3.74**	**3.88**	**4.59**	**4.26**	**4.51**
CHEESE:										
Natural		4.45	4.55	4.52	3.75	3.49	3.41	4.18	3.29	3.49
Processed		0.30	0.39	0.36	0.36	0.41	0.31	0.32	0.28	0.26
Total cheese		**4.75**	**4.94**	**4.88**	**4.11**	**3.89**	**3.72**	**4.50**	**3.58**	**3.75**
MEAT:										
Beef and veal		4.88	4.25	4.45	5.18	4.87	4.68	5.47	4.87	5.35
Mutton and lamb		2.42	2.48	2.46	2.19	2.24	2.30	3.58	2.77	3.94
Pork		1.98	2.01	2.00	2.53	2.64	2.37	2.82	2.61	2.56
Total carcase meat		9.28	8.74	8.91	9.90	9.76	9.35	11.87	10.24	11.85
Bacon and ham, uncooked		2.05	2.25	2.19	2.50	2.55	2.68	3.94	3.01	3.86
Poultry, uncooked		7.90	7.36	7.53	7.53	8.24	6.72	9.17	6.69	7.24
Other meat and meat products		10.94	11.06	11.02	12.94	13.48	13.64	12.95	14.42	12.86
Total meat		**30.17**	**29.41**	**29.65**	**32.87**	**34.03**	**32.39**	**37.93**	**34.36**	**35.81**
FISH:										
Fresh		1.48	1.22	1.30	0.85	0.73	0.69	2.79	1.03	1.96
Processed and shell		1.03	0.57	0.72	0.60	0.39	0.31	1.35	0.55	0.70
Prepared, including fish products		1.55	1.85	1.76	1.95	1.85	1.83	2.49	1.82	2.09
Frozen, including fish products		1.01	1.43	1.30	1.35	1.49	1.30	1.89	1.92	1.85
Total fish		**5.07**	**5.08**	**5.08**	**4.74**	**4.46**	**4.13**	**8.53**	**5.33**	**6.61**
EGGS	(no)	1.43	1.54	1.51	1.75	2.06	2.38	2.57	2.53	2.91
(Eggs purchased)	(no)	1.37	1.52	1.47	1.71	1.98	2.30	2.54	2.49	2.86
FATS:										
Butter		1.74	1.63	1.67	1.31	1.18	1.28	2.73	1.34	2.31
Margarine		1.70	2.57	2.29	2.52	2.62	2.85	3.71	3.09	4.10
Low fat and dairy spreads		1.06	1.47	1.35	1.83	1.89	1.65	1.87	1.70	2.11
Vegetable and salad oils	(fl oz)	1.64	1.71	1.69	1.73	1.71	2.29	1.77	1.61	1.29
Other fats		0.40	0.65	0.57	0.29	0.79	0.86	1.25	1.07	1.91
Total fats		**6.53**	**8.04**	**7.57**	**8.11**	**8.19**	**8.94**	**11.33**	**8.81**	**11.72**
SUGAR AND PRESERVES:										
Sugar		4.15	3.19	3.49	4.26	4.98	6.32	8.66	6.78	10.20
Honey, preserves, syrup and treacle		1.74	1.44	1.53	1.40	1.31	1.40	3.13	1.74	2.68
Total sugar and preserves		**5.89**	**4.63**	**5.02**	**5.66**	**6.29**	**7.72**	**11.79**	**8.52**	**12.88**
VEGETABLES:										
Potatoes		17.68	21.85	20.54	26.76	33.13	36.18	35.04	39.30	41.11
Fresh green		9.92	8.04	8.63	7.89	8.21	7.57	14.08	9.08	13.23
Other fresh		21.15	19.02	19.69	16.83	15.36	14.69	23.25	16.31	18.11
Frozen, including vegetable products		5.67	5.64	5.65	7.33	7.38	7.21	7.65	6.89	5.20
Other processed, including vegetable products		12.30	11.37	11.66	13.58	13.65	14.17	9.56	14.23	8.69
Total vegetables		**66.72**	**65.92**	**66.17**	**72.39**	**77.73**	**79.83**	**89.58**	**85.81**	**86.33**

Table B5 *continued*

ounces per person per week, except where otherwise stated

		Households with one or more earners						Households without an earner		OAP
		£730 and over	£520 and under £730	£520 and over	£280 and under £520	£140 and under £280	Less than £140	£140 or more	Less than £140	
		A1	A2	All A	B	C	D	E1	E2	
FRUIT:										
Fresh		29.61	26.95	27.79	22.97	18.22	16.93	37.12	18.38	26.61
Other, including fruit products		3.68	3.07	3.27	2.92	2.57	2.25	7.07	3.00	5.66
Fruit juices	(fl oz)	15.36	15.03	15.13	9.34	6.28	5.13	10.00	5.46	5.18
Total fruit		**48.65**	**45.05**	**46.19**	**35.23**	**27.07**	**24.31**	**54.19**	**26.84**	**37.45**
CEREALS:										
White bread (standard loaves)		6.58	6.20	6.32	11.04	14.56	16.64	9.48	16.31	14.21
Softgrain and premium loaves		2.72	1.97	2.20	2.38	2.17	2.23	1.41	2.14	2.00
Brown bread		3.19	3.51	3.41	3.01	3.16	3.04	3.57	3.61	4.52
Wholemeal bread		3.28	5.11	4.53	3.68	3.15	3.42	7.12	3.52	5.72
Other bread		5.62	5.29	5.39	4.68	4.11	3.81	4.94	3.42	4.19
Total bread		21.39	22.07	21.85	24.80	27.15	29.13	26.52	28.98	30.65
Flour		1.21	3.15	2.54	2.21	2.43	3.78	4.62	3.06	4.76
Cakes		3.55	4.09	3.91	3.86	4.02	3.51	5.68	3.78	5.62
Biscuits		5.13	4.52	4.71	5.12	5.30	4.96	5.82	5.06	6.03
Oatmeal and oat products		0.51	0.48	0.49	0.42	0.38	0.52	0.75	0.80	1.09
Breakfast cereals		5.20	5.49	5.40	4.90	4.30	3.75	5.49	4.98	4.53
Other cereals		9.47	9.10	9.14	8.17	7.78	7.54	6.03	7.28	4.43
Total cereals		**46.45**	**48.90**	**48.13**	**49.49**	**51.36**	**53.20**	**54.91**	**53.94**	**57.11**
BEVERAGES:										
Tea		1.14	1.08	1.10	1.09	1.26	1.27	2.11	1.61	2.56
Coffee		0.91	0.70	0.77	0.64	0.60	0.51	1.00	0.54	0.65
Cocoa and drinking chocolate		0.15	0.09	0.11	0.11	0.11	0.13	0.11	0.15	0.15
Branded food drinks		0.14	0.28	0.23	0.20	0.20	0.14	0.33	0.31	0.44
Total beverages		**2.34**	**2.15**	**2.21**	**2.04**	**2.17**	**2.04**	**3.55**	**2.61**	**3.82**
MISCELLANEOUS:										
Soups, canned, dehydrated and powdered		2.52	2.39	2.43	2.41	2.43	2.70	3.43	2.82	2.97
Other foods		42.15	14.31	23.06	11.67	9.28	8.94	13.42	8.21	7.37
Total miscellaneous		**44.67**	**16.70**	**25.49**	**14.08**	**11.71**	**11.64**	**16.85**	**11.03**	**10.34**
SOFT DRINKS:										
Concentrated	(fl oz)	3.94	6.28	5.55	4.59	4.61	4.77	2.85	4.57	2.50
Unconcentrated	(fl oz)	13.51	14.16	13.96	14.95	15.12	14.02	12.86	12.36	9.49
Low calorie, volume as purchased	(fl oz)	10.55	8.96	9.46	8.93	6.93	4.36	5.24	4.59	2.70
Total soft drinks, volume as purchased	(fl oz)	**28.00**	**29.40**	**28.96**	**28.47**	**26.66**	**23.14**	**20.95**	**21.52**	**14.69**
ALCOHOLIC DRINKS:										
Lager and beer	(cl)	16.32	17.59	17.19	22.26	17.42	13.06	11.58	11.05	6.96
Wine	(cl)	20.82	18.13	18.98	11.44	7.02	4.80	13.64	3.80	2.47
Other	(cl)	5.14	6.23	5.89	4.84	4.52	3.78	16.02	5.34	4.67
Total alcoholic drinks	(cl)	**42.28**	**41.95**	**42.05**	**38.54**	**28.97**	**21.64**	**41.23**	**20.18**	**14.10**
CONFECTIONERY:										
Chocolate confectionery	(g)	41.36	39.61	40.16	39.65	36.28	26.68	36.17	27.59	25.18
Mints and boiled sweets	(g)	7.86	10.16	9.44	12.88	12.54	15.75	17.85	15.91	19.34
Other	(g)	2.11	2.71	2.52	1.99	1.78	1.38	1.47	3.05	3.87
Total confectionery	(g)	**51.34**	**52.49**	**52.13**	**54.52**	**50.61**	**43.82**	**55.50**	**46.55**	**48.39**

Table B6

Household food expenditure on main food groups by income group, 1992

pence per person per week

	\multicolumn{6}{c}{Households with one or more earners}	\multicolumn{2}{c}{Households without an earner}	OAP
	£730 and over	£520 and under £730	£520 and over	£280 and under £520	£140 and under £280	Less than £140	£140 or more	Less than £140	
	A1	A2	All A	B	C	D	E1	E2	
MILK AND CREAM:									
Liquid wholemilk, full price	38.02	40.66	39.83	43.50	54.12	57.39	59.60	63.39	80.58
Welfare and school milk	0.10	0.13	0.12	0.24	0.20	0.31	0.04	0.28	...
Low fat milks	74.54	65.79	68.54	57.67	47.69	44.82	64.99	43.28	48.53
Dried and other milk	55.60	48.60	50.80	38.68	31.25	23.34	39.23	23.70	23.53
Cream	8.58	6.78	7.35	4.58	3.48	2.55	8.47	2.48	4.61
Total milk and cream	**176.86**	**161.96**	**166.64**	**144.68**	**136.73**	**128.41**	**172.33**	**133.13**	**157.25**
CHEESE:									
Natural	61.42	56.91	58.33	43.40	37.59	36.17	52.21	36.33	38.92
Processed	4.56	4.57	4.56	4.62	5.10	4.23	4.06	3.71	3.51
Total cheese	**65.98**	**61.48**	**62.89**	**48.02**	**42.69**	**40.40**	**56.27**	**40.04**	**42.43**
MEAT:									
Beef and veal	72.89	58.77	63.21	66.27	59.00	53.95	77.85	55.45	70.03
Mutton and lamb	35.35	28.86	30.89	24.40	23.08	22.81	40.82	27.26	39.68
Pork	23.42	24.39	24.09	24.61	25.22	22.79	31.03	22.77	25.58
Total carcase meat	131.66	112.02	118.19	115.28	107.30	99.55	149.70	105.48	135.30
Bacon and ham, uncooked	27.55	29.12	28.63	30.62	29.37	29.26	49.44	33.55	43.42
Poultry, uncooked	72.92	63.71	66.60	51.97	50.51	37.70	62.89	38.69	45.45
Other meat and meat products	172.23	146.11	154.31	158.28	144.14	123.85	140.05	126.39	127.37
Total meat	**404.36**	**350.96**	**367.73**	**356.15**	**331.32**	**290.36**	**402.08**	**304.11**	**351.54**
FISH:									
Fresh	30.32	20.11	23.32	12.80	10.21	9.23	45.72	13.91	28.10
Processed and shell	31.16	15.82	20.64	111.87	6.78	4.16	24.00	7.87	10.77
Prepared, including fish products	24.13	27.45	26.41	25.76	23.67	22.70	32.51	23.45	28.05
Frozen, including fish products	13.98	19.20	17.56	15.86	16.36	13.89	26.56	19.98	25.46
Total fish	**99.60**	**82.59**	**87.93**	**66.30**	**57.01**	**49.97**	**128.78**	**65.21**	**92.39**
EGGS	**14.90**	**15.98**	**15.64**	**16.10**	**17.92**	**19.75**	**24.56**	**22.63**	**26.47**
FATS:									
Butter	12.95	11.35	11.85	9.22	8.09	8.68	19.69	9.12	16.14
Margarine	6.33	9.50	8.51	9.35	9.58	9.21	13.91	10.47	14.64
Low fat and dairy spreads	5.08	7.89	7.01	9.52	9.53	8.11	9.32	7.93	10.82
Vegetable and salad oils	7.24	6.08	6.44	5.46	4.33	4.85	6.29	4.21	3.15
Other fats	2.46	3.42	3.11	3.07	2.74	2.98	4.23	3.44	4.91
Total fats	**34.06**	**38.24**	**36.93**	**36.62**	**34.27**	**33.83**	**53.44**	**35.17**	**51.41**
SUGAR AND PRESERVES:									
Sugar	9.35	7.07	7.79	8.22	9.43	11.88	17.34	12.72	19.56
Honey, preserves, syrup and treacle	9.91	7.73	8.41	6.80	6.16	6.03	15.05	8.28	13.07
Total sugar and preserves	**19.26**	**14.80**	**16.20**	**15.02**	**15.59**	**17.91**	**32.39**	**21.00**	**32.63**
VEGETABLES:									
Potatoes	24.97	23.48	23.95	23.46	24.19	26.14	29.30	27.58	29.68
Fresh green	42.14	30.94	34.46	21.90	20.20	17.41	37.07	20.43	27.95
Other fresh	85.95	68.59	74.05	52.49	44.41	39.31	62.87	40.86	41.69
Frozen, including vegetable products	23.82	20.86	21.79	25.81	24.35	22.59	27.97	20.92	15.37
Other processed, including vegetable products	73.25	66.53	68.64	72.31	65.30	61.60	44.18	55.81	34.18
Total vegetables	**250.13**	**210.40**	**222.89**	**195.97**	**178.45**	**167.05**	**201.39**	**165.60**	**148.86**

Table B6 *continued*

pence per person per week

	\u00a3730 and over (A1)	\u00a3520 and under \u00a3730 (A2)	\u00a3520 and over (All A)	\u00a3280 and under \u00a3520 (B)	\u00a3140 and under \u00a3280 (C)	Less than \u00a3140 (D)	\u00a3140 or more (E1)	Less than \u00a3140 (E2)	OAP
FRUIT:									
Fresh	111.72	95.31	100.46	74.10	56.63	51.96	117.32	56.39	77.65
Other, including fruit products	29.89	21.27	23.98	16.48	13.00	10.90	36.53	13.33	24.96
Fruit juices	42.51	36.34	38.28	21.72	14.56	11.22	23.78	12.24	11.47
Total fruit	**184.12**	**152.92**	**162.72**	**112.30**	**84.19**	**74.08**	**177.63**	**81.96**	**114.09**
CEREALS:									
White bread (standard loaves)	13.03	12.83	12.89	20.97	26.35	29.41	21.37	29.91	31.66
Softgrain and premium loaves	5.75	4.14	4.64	5.09	4.63	4.65	3.13	4.76	4.44
Brown bread	9.06	9.39	9.29	7.58	7.72	7.21	10.17	8.82	12.33
Wholemeal bread	7.77	11.93	10.62	8.71	7.35	7.77	19.19	8.99	15.65
Other bread	34.01	30.28	31.45	24.70	21.41	17.76	24.84	18.07	19.29
Total bread	69.61	68.56	68.89	67.04	67.46	66.79	78.69	70.55	83.37
Flour	1.87	3.77	3.18	2.96	3.20	4.91	6.32	3.99	6.22
Cakes	35.36	37.14	35.94	30.91	29.53	24.74	45.06	26.06	40.50
Biscuits	38.68	32.75	35.25	35.67	33.68	30.61	38.88	29.48	35.03
Oatmeal and oat products	2.40	1.73	1.94	1.45	1.19	1.30	2.86	2.05	3.27
Breakfast cereals	36.64	37.76	37.40	34.00	29.31	25.48	36.74	31.75	29.80
Other cereals	88.87	72.75	77.85	63.64	53.82	45.60	39.10	44.50	24.29
Total cereals	**273.43**	**254.46**	**260.42**	**235.67**	**218.18**	**199.49**	**247.65**	**208.38**	**222.48**
BEVERAGES:									
Tea	18.53	16.98	17.46	15.57	18.12	16.34	28.80	22.33	36.66
Coffee	27.00	19.54	21.88	19.71	18.72	16.52	28.53	16.36	20.01
Cocoa and drinking chocolate	1.79	1.20	1.38	0.98	1.05	1.24	1.23	1.44	1.31
Branded food drinks	1.69	5.06	4.00	3.05	2.75	1.77	4.27	3.67	5.15
Total beverages	**49.00**	**42.78**	**44.73**	**39.31**	**40.64**	**35.87**	**62.83**	**43.80**	**63.14**
MISCELLANEOUS:									
Soups, canned, dehydrated and powdered	11.06	10.66	10.79	9.14	8.56	8.37	14.73	8.99	10.93
Other foods	117.05	68.11	83.47	59.09	49.88	34.15	67.95	45.33	41.26
Total miscellaneous	**128.11**	**78.77**	**94.26**	**68.23**	**58.44**	**53.21**	**82.68**	**54.32**	**52.19**
TOTAL FOOD	**\u00a317.00**	**\u00a314.65**	**\u00a315.39**	**\u00a313.34**	**\u00a312.15**	**\u00a311.10**	**\u00a316.42**	**\u00a311.75**	**\u00a313.55**
SOFT DRINKS:									
Concentrated	12.24	17.46	15.82	12.47	11.96	12.05	7.47	11.33	6.97
Unconcentrated	24.22	26.17	25.56	23.09	21.46	19.20	20.25	17.04	13.27
Low calorie	16.82	15.73	16.08	13.65	10.62	6.60	8.26	6.17	3.34
Total soft drinks	**53.29**	**59.37**	**57.46**	**49.21**	**44.04**	**37.85**	**35.98**	**34.54**	**23.58**
ALCOHOLIC DRINKS:									
Lager and beer	27.31	29.27	28.66	34.05	25.14	17.29	15.13	14.94	10.29
Wine	95.47	75.39	81.70	44.49	24.56	15.79	49.48	15.04	11.42
Other	38.80	53.75	49.06	24.85	24.23	17.12	126.88	38.93	36.37
Total alcoholic drinks	**161.58**	**158.42**	**159.41**	**103.39**	**73.93**	**50.20**	**191.49**	**68.91**	**58.08**
CONFECTIONERY:									
Chocolate confectionery	24.60	19.04	20.79	19.71	17.28	13.20	19.26	12.57	12.77
Mints and boiled sweets	4.36	4.03	4.13	5.14	4.99	5.87	7.22	6.32	7.27
Other	1.15	1.21	1.19	0.94	0.89	0.63	0.57	1.27	1.36
Total confectionery	**30.11**	**24.28**	**26.11**	**25.78**	**23.16**	**19.69**	**27.06**	**20.16**	**21.40**
TOTAL FOOD AND DRINK	**\u00a319.45**	**\u00a317.07**	**\u00a317.82**	**\u00a315.13**	**\u00a313.57**	**\u00a312.18**	**\u00a318.97**	**\u00a312.99**	**\u00a314.58**

Table B7
Household food expenditure on main food groups by household composition, 1992

pence per person per week

	\multicolumn{10}{c}{Households with}	
No of adults	1		2					3	3 or more		4 or more	All households
No of children	0	1 or more	0	1	2	3	4 or more	0	1 or 2	3 or more	0	
MILK AND CREAM:												
Liquid wholemilk, full price	70.13	51.61	53.29	55.85	49.34	52.44	47.32	51.01	46.88	67.61	45.94	53.23
Welfare and school milk	0.01	0.68	–	0.12	0.48	0.49	0.52	–	0.12	0.81	–	0.21
Low fat milks	55.93	39.41	59.69	49.22	48.21	46.99	42.26	60.07	50.99	34.88	51.56	52.52
Dried and other milk	33.82	22.18	35.36	42.62	35.94	31.47	30.88	28.92	28.31	21.16	23.78	33.34
Cream	5.15	1.51	6.00	3.72	3.08	2.28	0.98	6.88	2.87	2.33	4.02	4.25
Total milk and cream	**165.04**	**115.39**	**154.36**	**151.52**	**137.04**	**133.67**	**121.96**	**146.87**	**129.16**	**126.77**	**125.29**	**143.53**
CHEESE:												
Natural	50.76	27.92	51.05	39.57	36.41	27.66	26.39	47.28	39.31	29.86	34.51	41.50
Processed	5.36	4.31	4.11	4.17	4.56	4.61	5.43	5.45	4.13	3.76	4.69	4.52
Total cheese	**56.12**	**32.22**	**55.17**	**43.74**	**40.98**	**32.27**	**31.82**	**52.73**	**43.43**	**33.63**	**39.21**	**46.03**
MEAT:												
Beef and veal	61.00	41.05	81.57	60.63	50.16	35.25	36.42	78.28	62.55	25.73	71.13	62.40
Mutton and lamb	31.60	13.62	38.94	20.36	16.85	16.23	10.79	31.71	26.15	27.43	28.61	26.54
Pork	23.53	17.21	29.76	21.32	20.10	16.94	19.36	30.37	29.56	12.96	34.54	24.75
Total carcase meat	116.14	71.89	150.27	102.31	87.11	68.42	66.57	140.35	118.25	66.12	134.28	113.69
Bacon and ham, uncooked	38.34	21.90	44.22	29.60	20.64	21.18	12.92	42.19	28.01	17.55	33.19	32.05
Poultry, uncooked	48.74	34.99	62.83	45.70	39.48	34.37	41.43	65.87	49.11	44.92	50.60	50.01
Other meat and meat products	169.53	114.64	166.72	152.59	125.38	108.06	112.81	157.43	134.61	110.38	128.35	144.11
Total meat	**372.74**	**243.41**	**424.04**	**330.19**	**272.61**	**232.02**	**233.73**	**405.84**	**329.99**	**238.97**	**346.41**	**339.86**
FISH:												
Fresh	20.85	4.64	28.79	9.48	8.13	5.08	5.33	16.79	8.42	5.09	13.80	15.25
Processed and shell	15.98	3.43	16.87	9.36	8.08	3.11	3.66	9.66	6.18	8.37	4.85	10.37
Prepared, including fish products	36.17	17.19	33.20	21.53	17.42	16.78	12.29	34.72	20.46	11.75	19.98	25.14
Frozen, including fish products	26.15	13.62	21.17	15.92	14.90	13.43	9.62	18.00	16.52	10.83	12.90	17.57
Total fish	**99.15**	**38.88**	**100.03**	**56.29**	**48.54**	**38.39**	**30.88**	**79.16**	**51.57**	**36.04**	**51.53**	**68.32**
EGGS	**27.34**	**16.59**	**23.22**	**16.51**	**14.49**	**12.88**	**12.44**	**20.93**	**14.97**	**13.47**	**18.97**	**18.77**
FATS:												
Butter	13.93	5.00	14.06	7.33	6.56	5.93	5.69	13.98	7.94	4.94	11.72	10.02
Margarine	13.19	7.69	11.53	8.65	8.37	7.95	11.55	10.42	9.69	9.86	10.87	10.05
Low fat and dairy spreads	10.91	6.84	10.80	8.86	7.93	7.62	3.91	9.96	9.48	4.01	8.39	9.12
Vegetable and salad oils	4.87	4.15	5.59	3.86	4.35	4.02	3.56	4.81	4.73	8.84	7.38	4.88
Other fats	3.67	1.56	5.04	2.31	2.24	1.91	1.97	4.16	3.09	1.60	4.00	3.33
Total fats	**46.56**	**25.25**	**47.01**	**31.00**	**29.45**	**27.43**	**26.67**	**43.34**	**34.93**	**29.26**	**42.37**	**37.40**
SUGAR AND PRESERVES:												
Sugar	13.74	10.09	13.91	8.94	6.64	7.24	8.06	11.55	10.50	12.23	11.30	10.61
Honey, preserves, syrup and treacle	13.46	3.69	10.53	5.78	4.83	4.94	4.57	9.75	5.79	4.34	6.11	7.66
Total sugar and preserves	**27.20**	**13.78**	**24.44**	**14.73**	**11.47**	**12.18**	**12.62**	**21.30**	**16.29**	**16.57**	**17.41**	**18.27**
VEGETABLES:												
Potatoes	27.82	23.22	29.86	23.73	21.60	16.40	18.75	28.55	23.17	21.66	28.06	25.12
Fresh green	32.28	12.43	32.59	18.87	16.06	13.90	13.38	28.09	17.48	13.15	20.41	22.95
Other fresh	61.63	30.62	65.53	48.77	37.13	31.75	26.53	54.71	42.55	31.13	43.11	49.01
Frozen, including vegetable products	21.26	20.52	25.65	24.96	22.16	20.25	23.98	26.62	23.04	26.92	20.63	23.56
Other processed, including vegetable products	58.45	64.78	61.70	72.17	64.48	57.71	65.28	64.89	61.64	57.66	59.62	63.11
Total vegetables	**201.43**	**151.57**	**215.40**	**188.51**	**161.43**	**140.01**	**147.93**	**202.86**	**167.88**	**150.51**	**171.84**	**183.75**

Table B7 *continued*

pence per person per week

	Households with											
No of adults	1		2					3	3 or more		4 or more	All households
No of children	0	1 or more	0	1	2	3	4 or more	0	1 or 2	3 or more	0	
FRUIT:												
Fresh	96.19	43.78	91.52	55.83	58.37	49.95	44.23	74.76	55.75	44.33	55.82	69.28
Other, including fruit products	22.77	6.07	24.71	12.65	12.37	10.51	9.63	19.46	12.22	17.31	13.10	16.72
Fruit juices	22.61	11.91	20.50	20.41	17.04	12.82	17.94	18.06	17.16	10.57	17.16	18.26
Total fruit	**141.56**	**61.76**	**136.73**	**88.89**	**87.78**	**73.28**	**71.80**	**112.29**	**85.12**	**72.20**	**86.08**	**104.26**
CEREALS:												
White bread (standard loaves)	26.74	23.79	26.61	22.80	20.88	23.55	25.75	26.60	25.57	18.21	24.07	24.46
Softgrain and premium loaves	4.98	5.35	5.06	4.84	4.86	3.34	2.64	3.81	4.70	3.24	6.04	4.70
Brown bread	12.98	4.84	12.24	6.86	4.80	4.13	3.62	9.82	6.97	7.73	7.12	8.32
Wholemeal bread	16.12	4.82	13.01	8.04	6.19	5.00	6.43	11.66	6.29	5.45	7.33	9.39
Other bread	28.37	14.87	26.63	23.21	19.98	15.73	16.85	26.45	18.37	11.05	21.57	22.52
Total bread	89.19	53.67	83.55	65.76	56.72	51.75	55.29	78.35	61.91	45.67	66.12	69.40
Flour	3.87	4.12	4.63	2.51	2.17	3.58	2.42	3.47	3.43	9.11	7.74	3.74
Cakes	42.18	19.50	38.98	28.76	25.14	21.21	23.78	37.73	24.65	20.19	29.97	31.18
Biscuits	38.32	30.80	34.73	34.15	36.00	35.12	31.46	33.31	29.90	24.06	27.34	34.04
Oatmeal and oat products	3.28	1.22	2.14	1.28	1.09	0.89	1.01	1.92	1.18	0.82	1.53	1.66
Breakfast cereals	35.00	32.93	30.85	29.72	34.53	34.17	34.05	27.38	29.31	27.08	28.57	31.63
Other cereals	49.22	51.30	53.79	62.00	57.45	49.71	55.18	51.03	52.69	48.48	52.55	54.11
Total cereals	**261.06**	**193.54**	**248.68**	**224.19**	**213.10**	**196.43**	**203.19**	**233.20**	**203.07**	**175.41**	**213.83**	**225.75**
BEVERAGES:												
Tea	28.36	13.85	26.22	15.55	11.07	11.60	16.49	24.17	18.15	12.40	22.39	19.42
Coffee	26.06	15.63	25.05	16.42	14.03	12.48	10.96	23.05	18.65	14.37	17.83	19.36
Cocoa and drinking chocolate	1.10	0.91	1.32	1.25	0.79	1.28	1.90	1.37	0.98	1.24	0.80	1.14
Branded food drinks	4.49	1.84	4.36	2.63	1.51	2.18	3.31	4.79	2.13	2.86	4.08	3.19
Total beverages	**60.01**	**32.23**	**56.95**	**35.85**	**27.39**	**27.53**	**32.65**	**53.30**	**39.90**	**30.86**	**45.10**	**43.11**
MISCELLANEOUS:												
Soups, dehydrated, canned and powdered	14.02	7.11	11.71	7.73	7.36	7.14	5.18	10.29	8.16	7.35	8.61	9.39
Other foods	53.93	36.55	63.67	57.21	49.56	42.52	42.98	67.40	48.12	35.69	54.87	54.59
Total miscellaneous	**67.95**	**43.66**	**75.38**	**64.94**	**56.92**	**49.66**	**48.16**	**77.69**	**56.28**	**43.04**	**63.47**	**63.98**
TOTAL FOOD	**£15.26**	**£9.68**	**£15.61**	**£12.46**	**£11.01**	**£9.76**	**£9.74**	**£14.50**	**£11.73**	**£9.66**	**£12.22**	**£12.93**
SOFT DRINKS:												
Concentrated	7.99	17.11	7.35	11.83	16.71	18.36	21.25	8.76	11.97	11.44	8.57	11.79
Unconcentrated	20.57	23.10	18.20	24.27	22.17	20.57	16.72	19.49	24.32	31.86	18.76	20.95
Low calorie	8.41	7.63	11.54	12.53	11.83	8.56	10.83	9.53	9.01	6.26	9.73	10.49
Total soft drinks	**36.97**	**47.82**	**37.09**	**48.64**	**50.71**	**47.49**	**48.80**	**37.78**	**45.30**	**49.57**	**37.06**	**43.24**
ALCOHOLIC DRINKS:												
Lager and beer	34.08	7.31	33.48	27.58	20.67	15.19	16.81	27.22	16.81	3.02	20.80	24.80
Wine	42.94	14.22	55.75	35.02	23.80	13.14	6.04	32.89	17.84	8.76	26.16	33.52
Other	52.94	7.84	62.14	19.25	16.38	6.16	9.15	38.34	16.63	5.21	28.15	32.66
Total alcoholic drinks	**129.95**	**29.38**	**151.37**	**81.86**	**60.84**	**34.48**	**31.98**	**98.44**	**51.28**	**16.99**	**75.11**	**90.98**
CONFECTIONERY:												
Chocolate confectionery	15.71	13.10	17.15	21.42	18.05	23.47	13.37	14.89	16.00	9.92	10.72	17.19
Mints and boiled sweets	5.99	5.28	6.81	4.47	5.28	5.27	4.43	4.71	4.84	3.13	3.77	5.46
Other	1.55	0.55	1.15	0.65	0.95	1.04	0.78	0.97	0.61	0.24	0.76	0.96
Total confectionery	**23.25**	**18.92**	**25.11**	**26.53**	**24.27**	**29.78**	**18.58**	**20.58**	**21.44**	**13.28**	**15.25**	**23.61**
TOTAL FOOD AND DRINK	**£17.16**	**£10.64**	**£17.75**	**£14.03**	**£12.37**	**£10.88**	**£10.73**	**£16.06**	**£12.91**	**£10.47**	**£13.49**	**£14.51**

Table B8

Household food consumption by household composition groups within income groups: selected food items, 1992

ounces per person per week[a]

Income group A

Households[b] with

		Adults only	2 adults and 1 child	2 adults and 2 children	2 adults and 3 children	3 or more adults, 1 or more children
Milk and cream	(pt or eq pt)	3.77	4.23	3.78	4.41	3.57
Cheese		5.56	4.74	4.19	4.83	4.23
Carcase meat		11.48	8.21	7.33	8.31	7.74
Other meat and meat products		23.30	21.09	17.06	23.04	19.96
Fish		6.92	4.89	4.55	3.13	4.06
Eggs	(no)	1.84	1.64	1.35	1.38	1.01
Fats		8.13	6.17	6.86	7.64	8.55
Sugar and preserves		6.62	4.04	3.08	5.10	6.21
Potatoes		21.35	14.95	23.57	16.76	22.85
Fresh green vegetables		12.31	8.34	6.78	5.53	7.13
Other fresh vegetables		24.41	18.03	16.06	14.94	18.71
Processed vegetables		17.31	17.74	14.40	21.24	15.15
Fresh fruit		36.01	21.08	22.66	26.85	25.57
Other fruit and fruit products		15.71	19.43	16.15	17.01	19.32
Bread		24.63	23.91	18.32	19.75	20.90
Other cereals		24.05	24.53	22.08	39.17	23.95
Tea		1.45	1.04	0.67	0.79	1.28
Coffee		0.84	0.97	0.40	1.01	0.93
Cocoa and drinking chocolate		0.11	–	–	0.30	0.09
Branded food drinks		0.22	0.28	0.04	0.11	0.30
FOOD EXPENDITURE		**£18.75**	**£14.98**	**£12.40**	**£14.38**	**£14.50**
Soft drinks	(fl oz)	21.45	35.05	28.81	32.62	34.32
Alcoholic drinks	(cl)	71.47	49.40	21.23	18.48	28.38
Confectionery	(g)	44.59	62.58	49.27	83.73	35.28
TOTAL EXPENDITURE		**£22.50**	**£17.53**	**£14.03**	**£16.06**	**£16.13**

Income group B

Households with

		Adults only	1 adult, 1 or more children[c]	2 adults and 1 child	2 adults and 2 children	2 adults and 3 children	2 adults and 4 or more children	3 or more adults, 1 or more children
Milk and cream	(pt or eq pt)	3.74	4.21	4.03	3.75	3.54	3.51	3.49
Cheese		4.96	3.68	3.48	3.93	2.59	2.98	3.97
Carcase meat		11.87	9.64	8.48	8.13	6.64	8.95	12.01
Other meat and meat products		25.89	22.55	23.32	19.97	19.11	21.38	23.17
Fish		6.17	3.07	4.15	4.15	3.73	2.95	3.35
Eggs	(no)	2.01	1.05	1.74	1.62	1.53	1.12	1.53
Fats		9.48	7.28	6.53	7.15	6.76	5.72	9.38
Sugar and preserves		6.47	5.05	6.01	4.36	4.38	4.48	6.88
Potatoes		30.35	21.84	24.14	25.27	19.20	22.16	28.99
Fresh green vegetables		10.07	4.34	7.31	6.49	5.92	4.59	7.13
Other fresh vegetables		21.30	11.91	16.09	13.96	11.29	9.12	15.99
Processed vegetables		22.12	26.23	21.90	20.08	17.10	20.57	20.34
Fresh fruit		27.86	15.57	21.32	21.52	17.28	15.23	18.84
Other fruit and fruit products		14.77	21.73	12.61	10.65	8.72	5.75	11.00
Bread		26.64	19.59	23.34	23.64	22.93	25.23	24.73
Other cereals		25.16	27.05	24.96	25.91	22.62	24.94	21.36
Tea		1.33	1.11	1.10	0.81	0.75	1.16	1.17
Coffee		0.91	0.86	0.49	0.45	0.41	0.55	0.50
Cocoa and drinking chocolate		0.13	–	0.09	0.08	0.12	0.16	0.13
Branded food drinks		0.23	0.20	0.24	0.13	0.27	0.04	0.15
FOOD EXPENDITURE		**£15.74**	**£12.60**	**£13.24**	**£11.99**	**£9.89**	**£10.28**	**£12.11**
Soft drinks	(fl oz)	25.09	28.11	28.84	30.60	28.24	32.26	33.98
Alcoholic drinks	(cl)	55.50	–	40.57	28.79	23.31	31.20	17.40
Confectionery	(g)	49.23	44.50	54.48	61.45	67.20	40.20	50.62
TOTAL EXPENDITURE		**£18.08**	**£13.26**	**£14.96**	**£13.47**	**£11.19**	**£11.50**	**£13.39**

Table B8 *continued*

ounces per person per week [a]

	Income group C						
	Households with						
	Adults only	1 adult, 1 or more children	2 adults and				3 or more adults, 1 or more children
			1 child	2 children	3 children	4 or more children	
Milk and cream (pt or eq pt)	3.87	3.31	3.86	3.56	3.46	3.90	3.67
Cheese	4.48	3.72	4.19	3.39	2.18	2.25	3.78
Carcase meat	11.80	6.87	9.53	7.59	6.79	6.51	9.39
Other meat and meat products	28.48	24.61	23.76	20.70	17.02	20.27	21.03
Fish	5.53	4.69	3.92	3.48	2.57	3.48	4.23
Eggs (no)	2.46	1.56	2.04	1.71	1.37	1.89	1.74
Fats	9.40	5.42	7.68	7.26	6.82	7.76	7.46
Sugar and preserves	7.71	3.88	5.87	4.39	4.23	6.07	6.73
Potatoes	35.48	24.63	36.45	29.23	26.65	40.18	32.25
Fresh green vegetables	10.27	5.71	7.49	6.13	5.29	6.33	7.83
Other fresh vegetables	19.07	16.19	14.92	11.39	9.54	14.98	12.89
Processed vegetables	23.08	19.66	21.71	19.41	14.06	21.67	20.37
Fresh fruit	21.99	23.83	14.74	15.28	13.73	10.76	16.63
Other fruit and fruit products	11.06	7.22	8.62	7.66	3.62	4.02	7.67
Bread	30.22	22.45	27.44	24.06	23.29	23.41	24.97
Other cereals	25.72	27.88	23.45	22.37	22.43	19.41	24.14
Tea	1.57	1.05	1.03	0.87	0.88	1.35	1.29
Coffee	0.76	0.62	0.56	0.48	0.29	0.28	0.56
Cocoa and drinking chocolate	0.11	0.07	0.19	0.13	0.10	–	0.06
Branded food drinks	0.35	0.11	0.14	0.07	0.12	–	0.07
FOOD EXPENDITURE	**£14.52**	**£11.15**	**£12.01**	**£10.12**	**£8.30**	**£8.86**	**£10.48**
Soft drinks (fl oz)	25.20	31.23	29.42	30.24	24.98	27.62	22.70
Alcoholic drinks (cl)	39.14	15.95	27.23	24.28	11.44	7.80	19.39
Confectionery (g)	49.99	36.87	61.49	52.21	60.61	43.50	34.39
TOTAL EXPENDITURE	**£16.27**	**£12.46**	**£13.45**	**£11.39**	**£9.19**	**£9.59**	**£11.35**

	Income group D and E2						
	Households with						
	Adults only	1 adult, 1 or more children	2 adults and				3 or more adults, 1 or more children
			1 child	2 children	3 children	4 or more children [c]	
Milk and cream (pt or eq pt)	4.42	3.75	3.73	3.77	4.15	3.81	3.50
Cheese	4.42	2.66	3.43	3.04	3.03	2.23	3.39
Carcase meat	12.30	7.42	8.20	8.57	5.63	4.53	8.49
Other meat and meat products	26.43	20.20	22.49	20.41	19.68	23.70	23.57
Fish	6.44	3.35	3.19	2.81	3.88	2.66	3.44
Eggs (no)	2.88	2.07	2.01	2.03	2.07	1.92	2.38
Fats	10.35	6.56	7.58	7.33	7.89	11.39	8.92
Sugar and preserves	9.84	6.42	6.73	5.97	6.86	6.99	7.90
Potatoes	41.39	31.21	39.79	30.60	29.23	45.53	46.15
Fresh green vegetables	12.41	4.28	5.83	4.65	2.96	3.96	6.88
Other fresh vegetables	21.57	10.01	13.41	9.40	9.59	6.35	10.63
Processed vegetables	20.06	21.19	24.23	23.73	26.17	21.00	17.25
Fresh fruit	25.13	11.67	9.69	12.49	9.74	6.30	11.33
Other fruit and fruit products	10.74	5.51	5.16	7.04	5.62	1.24	3.94
Bread	32.91	23.85	25.60	23.23	25.82	34.99	31.59
Other cereals	28.32	22.50	18.93	21.39	19.47	18.13	22.45
Tea	1.85	1.02	1.21	0.88	1.36	0.95	1.33
Coffee	0.65	0.44	0.44	0.37	0.38	0.24	0.42
Cocoa and drinking chocolate	0.16	0.08	0.27	0.09	0.12	0.15	0.11
Branded food drinks	0.34	0.12	0.18	0.11	0.18	–	0.06
FOOD EXPENDITURE	**£14.17**	**£9.03**	**£9.71**	**£8.99**	**£8.55**	**£6.82**	**£9.54**
Soft drinks (fl oz)	19.70	27.54	22.44	24.61	24.23	19.08	21.90
Alcoholic drinks (cl)	27.97	9.55	17.91	22.12	11.32	16.92	12.17
Confectionery (g)	49.44	43.86	38.73	32.24	46.33	44.06	50.86
TOTAL EXPENDITURE	**£15.64**	**£9.94**	**£10.50**	**£9.97**	**£9.33**	**£7.41**	**£10.49**

(a) except where otherwise stated

(b) averages are not shown for households of 1 adult and 1 or more children or 2 adults with 4 or more children in income group A because there are fewer than 10 such households in the sample

(c) the figures in this column are based on a sample of more than 9 but fewer than 20 households

Table B9
Nutritional value of household food: national averages, 1990 - 1992

		1990	1991	1992	1992[a]
		(i) Consumption per person per day			
Energy	(kcal)	1,870	1,840	1,860	1,960
	(MJ)	7.9	7.8	7.8	8.2
Total protein	(g)	63.1	62.3	62.8	63.4
Animal protein	(g)	38.7	38.3	38.6	39.1
Fat	(g)	86	85	86	87
Fatty acids:					
saturated	(g)	34.6	33.7	33.6	34.3
monounsaturated	(g)	31.8	31.5	31.8	32.2
polyunsaturated	(g)	13.9	13.8	14.4	14.5
Carbohydrate[b]	(g)	224	223	222	240
of which:					
non-milk extrinsic sugars	(g)	57	58	53	70
Calcium	(mg)	820	810	830	840
Iron	(mg)	10.4	10.1	10.1	10.3
Zinc	(mg)	7.8	7.8	7.9	7.9
Magnesium	(mg)	222	217	221	232
Sodium	(g)	2.50	2.49	2.51	2.53
Potassium	(g)	2.48	2.48	2.49	2.54
Thiamin	(mg)	1.28	1.28	1.26	1.26
Riboflavin	(mg)	1.61	1.62	1.61	1.64
Niacin equivalent	(mg)	25.1	25.1	24.9	25.6
Vitamin B6	(mg)	1.6	1.6	1.7	1.8
Vitamin B12	(µg)	5.0	4.9	4.9	5.0
Folate	(µg)	244	239	243	249
Vitamin C	(mg)	52	55	51	54
Vitamin A:					
retinol	(µg)	780	800	860	860
β-carotene	(µg)	1,880	1,910	1,750	1,750
total (retinol equivalent)	(µg)	1,100	1,110	1,150	1,150
Vitamin D[c]	(µg)	3.02	3.12	2.97	2.97
	(ii) As a percentage of Reference Nutrient Intake[d]				
Energy[e]		91	89	89	93
Protein		142	139	139	140
Calcium		120	118	120	122
Iron		101	97	96	98
Zinc		100	99	99	99
Magnesium		84	84	84	88
Sodium		na	170	169	170
Potassium		na	80	79	80
Thiamin		154	153	150	150
Riboflavin		144	144	142	144
Niacin equivalent		183	182	179	184
Vitamin B6		133	132	139	146
Vitamin B12		369	360	354	361
Folate		131	129	130	133
Vitamin C		138	144	131	141
Vitamin A (retinol equivalent)		179	181	185	186
	(iii) As a percentage of food energy[a]				
Fat		41.6	41.4	41.7	40.1
of which:					
saturated fatty acids		16.6	16.4	16.3	15.8
Carbohydrate		44.9	45.3	44.8	45.9

(a) final column includes soft and alcoholic drinks, and confectionery

(b) available carbohydrate, calculated as monosaccharide

(c) contributions from pharmaceutical sources of this (or any other) vitamin are not recorded by the Survey

(d) Department of Health, *Dietary Reference Values for Food Energy and Nutrients for the United Kingdom,* HMSO, 1991. Before comparison with the Reference Nutrient Intakes ten percent has first been deducted from each absolute intake given above to allow for wastage, and an allowance has also been made for meals not taken from the domestic food supply

(e) as a percentage of Estimated Average Requirement

Table B10
Nutritional value of household food by region, 1992

		Scotland	Wales	England	North	Yorkshire and Humberside	North West	East Midlands	West Midlands	South West	South East/ East Anglia
					\multicolumn{7}{c}{Regions of England}		

		Scotland	Wales	England	North	Yorkshire and Humberside	North West	East Midlands	West Midlands	South West	South East/East Anglia
\multicolumn{12}{c}{(i) Consumption per person per day}
Energy	(kcal)	1,830	1,870	1,860	1,860	1,950	1,830	1,950	1,810	1,870	1,850
	(MJ)	7.7	7.8	7.8	7.8	8.2	7.7	8.2	7.6	7.8	7.8
Total protein	(g)	62.7	64.0	62.8	63.8	65.2	62.3	63.8	61.6	61.5	62.6
Animal protein	(g)	38.8	39.2	38.5	39.3	40.0	38.7	38.0	38.2	36.9	38.6
Fat	(g)	84	85	86	87	90	86	89	83	85	86
Fatty acids:											
saturated	(g)	33.5	33.2	33.6	33.9	35.4	33.2	34.0	32.8	33.5	33.6
monounsaturated	(g)	30.9	31.6	31.9	32.2	33.4	31.8	32.8	31.0	31.4	31.7
polyunsaturated	(g)	13.8	14.0	14.5	14.3	15.0	14.2	15.6	13.5	14.1	14.7
Carbohydrate	(g)	219	226	222	221	234	216	238	216	227	220
of which:											
non-milk extrinsic sugars	(g)	52	55	53	50	57	50	56	54	56	53
Calcium	(mg)	830	840	820	800	830	780	890	800	840	830
Iron	(mg)	10.1	10.1	10.1	10.3	10.7	9.8	10.3	9.5	10.1	10.1
Zinc	(mg)	7.9	7.9	7.9	8.1	8.2	7.8	7.9	7.5	7.7	7.9
Magnesium	(mg)	218	225	221	218	228	212	225	210	229	224
Sodium	(g)	2.59	2.55	2.50	2.64	2.66	2.55	2.51	2.52	2.43	2.44
Potassium	(g)	2.42	2.52	2.49	2.48	2.56	2.41	2.54	2.37	2.56	2.52
Thiamin	(mg)	1.23	1.29	1.26	1.27	1.32	1.23	1.32	1.21	1.30	1.25
Riboflavin	(mg)	1.59	1.66	1.61	1.58	1.65	1.56	1.63	1.54	1.64	1.63
Niacin equivalent	(mg)	24.6	25.1	24.9	25.1	26.0	24.8	25.0	24.2	24.4	25.0
Vitamin B6	(mg)	1.6	1.7	1.7	1.7	1.8	1.7	1.7	1.6	1.8	1.7
Vitamin B12	(µg)	4.9	4.8	4.9	4.9	5.2	4.9	4.8	4.7	4.7	4.9
Folate	(µg)	230	244	244	238	250	230	250	230	253	250
Vitamin C	(mg)	50	48	51	47	49	46	49	44	53	56
Vitamin A:											
retinol	(µg)	780	820	870	790	900	850	810	880	830	920
β-carotene	(µg)	1,670	1,870	1,750	1,820	1,800	1,670	1,780	1,610	1,780	1,770
total (retinol equivalent)	(µg)	1,060	1,130	1,160	1,090	1,200	1,130	1,100	1,150	1,130	1,200
Vitamin D	(µg)	2.87	3.17	2.96	2.84	3.11	3.07	3.03	2.80	3.05	2.93
\multicolumn{12}{c}{(ii) As a percentage of Reference Nutrient Intake[a]}
Energy[b]		88	89	89	90	93	89	92	85	87	88
Protein		139	140	139	142	144	141	141	134	133	138
Calcium		120	121	120	118	120	116	128	115	121	121
Iron		96	96	96	99	102	95	97	88	96	96
Zinc		100	99	99	103	103	100	100	93	95	99
Magnesium		83	85	84	84	86	83	85	79	85	85
Sodium		173	171	169	180	179	176	168	167	161	164
Potassium		76	79	79	80	81	78	80	74	80	80
Thiamin		147	154	150	152	157	149	156	141	151	149
Riboflavin		140	145	141	140	144	140	142	133	141	143
Niacin equivalent		178	181	179	182	186	181	179	171	172	180
Vitamin B6		135	141	140	141	145	138	140	131	140	141
Vitamin B12		355	350	354	358	378	361	344	337	334	359
Folate		122	129	130	128	133	125	133	121	133	134
Vitamin C		131	124	132	122	127	123	128	113	134	144
Vitamin A (retinol equivalent)		171	182	187	177	192	185	177	181	178	195
\multicolumn{12}{c}{(iii) As a percentage of food energy}
Fat		41.5	41.0	41.7	41.9	41.7	42.1	41.1	41.6	41.1	41.9
of which:											
saturated fatty acids		16.5	16.0	16.3	16.4	16.3	16.4	15.7	16.3	16.1	16.3
Carbohydrate		44.8	45.4	44.9	44.4	44.9	44.3	45.8	44.8	45.7	44.5
\multicolumn{12}{c}{(iv) Contributions to selected nutrients from soft and alcoholic drinks and confectionery}
Energy	(kcal)	110	90	100	100	100	100	90	100	100	110
	(MJ)	0.4	0.4	0.4	0.4	0.4	0.4	0.4	0.4	0.4	0.4
Fat	(g)	2	1	1	1	1	1	1	1	1	1
Carbohydrate	(g)	18	16	18	17	18	17	17	18	17	18

(a) Department of Health, *Dietary Reference Values for Food Energy and Nutrients for the United Kingdom*, HMSO, 1991
(b) as a percentage of Estimated Average Requirement

Table B11

Nutritional value of household food by income group, 1992

		\multicolumn{7}{c}{Income groups}
		\multicolumn{7}{c}{Gross weekly income of head of household}
		\multicolumn{4}{c}{Households with one or more earners}	\multicolumn{2}{c}{Households without an earner}	OAP
		£520 and over	£280 and under £520	£140 and under £280	Less than £140	£140 or more	Less than £140	
		A	B	C	D	E1	E2	
\multicolumn{9}{c}{(i) Consumption per person per day}
Energy	(kcal)	1,750	1,780	1,810	1,870	2,200	1,930	2,170
	(MJ)	7.3	7.5	7.6	7.8	9.2	8.1	9.1
Total protein	(g)	59.8	61.0	61.7	61.9	73.1	65.2	69.9
Animal protein	(g)	36.7	37.5	37.7	37.2	46.5	39.9	43.5
Fat	(g)	81	83	84	86	102	88	100
Fatty acids:								
saturated	(g)	32.3	32.2	32.5	32.9	40.8	34.6	40.5
monounsaturated	(g)	29.5	30.7	31.0	32.0	37.2	32.8	36.7
polyunsaturated	(g)	13.9	14.3	14.1	14.8	16.3	14.5	15.2
Carbohydrate	(g)	207	211	216	225	265	234	263
of which:								
non-milk extrinsic sugars	(g)	49	49	49	52	78	57	77
Calcium	(mg)	820	800	800	820	960	860	920
Iron	(mg)	9.9	9.9	9.8	9.7	11.6	10.5	11.0
Zinc	(mg)	7.7	7.7	7.7	7.7	9.2	8.2	8.7
Magnesium	(mg)	218	215	211	215	271	234	253
Sodium	(g)	2.35	2.46	2.50	2.52	2.74	2.61	2.67
Potassium	(g)	2.41	2.41	2.41	2.42	3.03	2.59	2.82
Thiamin	(mg)	1.21	1.22	1.22	1.23	1.48	1.33	1.44
Riboflavin	(mg)	1.59	1.55	1.55	1.53	1.91	1.72	1.87
Niacin equivalent	(mg)	23.9	24.3	24.5	24.1	29.2	25.7	27.3
Vitamin B6	(mg)	1.7	1.7	1.7	1.6	2.0	1.8	1.9
Vitamin B12	(µg)	4.5	4.6	4.7	4.7	6.0	5.4	5.8
Folate	(µg)	239	234	235	234	295	257	280
Vitamin C	(mg)	62	52	46	44	71	49	52
Vitamin A:								
retinol	(µg)	700	770	850	820	970	1,040	1,220
β-carotene	(µg)	1,830	1,750	1,690	1,570	2,200	1,790	1,780
total (retinol equivalent)	(µg)	1,010	1,060	1,130	1,080	1,330	1,330	1,520
Vitamin D	(µg)	2.62	2.81	2.75	2.79	4.52	3.23	3.82
\multicolumn{9}{c}{(ii) As a percentage of Reference Nutrient Intake[a]}
Energy[b]		88	86	86	88	100	93	97
Protein		143	140	137	136	145	144	131
Calcium		126	119	117	118	131	124	121
Iron		95	94	91	90	120	102	114
Zinc		102	98	97	95	108	103	101
Magnesium		89	85	81	81	93	88	83
Sodium		170	172	170	168	167	175	153
Potassium		83	80	77	76	85	82	74
Thiamin		152	147	146	144	168	158	157
Riboflavin		149	140	137	133	155	150	146
Niacin equivalent		181	177	176	170	201	186	185
Vitamin B6		145	140	136	132	153	146	136
Vitamin B12		352	350	345	340	392	388	354
Folate		137	129	126	124	143	136	129
Vitamin C		172	140	119	111	171	123	119
Vitamin A (retinol equivalent)		173	175	183	171	199	211	219
\multicolumn{9}{c}{(iii) As a percentage of food energy}
Fat		41.9	42.0	41.6	41.5	41.6	41.1	41.5
of which:								
Saturated fatty acids		16.6	16.2	16.1	15.9	16.7	16.1	16.8
Carbohydrate		44.4	44.4	44.8	45.2	45.1	45.4	45.6
\multicolumn{9}{c}{(iv) Contributions to selected nutrients from soft and alcoholic drinks and confectionery}
Energy	(kcal)	120	110	100	90	120	90	70
	(MJ)	0.5	0.5	0.4	0.4	0.5	0.4	0.3
Fat	(g)	1	1	1	1	1	1	1
Carbohydrate	(g)	19	19	18	17	15	16	12

(a) Department of Health, *Dietary Reference Values for Food Energy and Nutrients for the United Kingdom,* HMSO, 1991
(b) as a percentage of Estimated Average Requirement

Table B12

Nutritional value of household food by household composition, 1992

		\multicolumn{10}{c}{Households with}
No of adults		1		2					3	3 or more		4 or more
No of children		0	1 or more	0	1	2	3	4 or more	0	1 or 2	3 or more	0

(i) Consumption per person per day

Energy	(kcal)	2,110	1,630	2,110	1,740	1,630	1,590	1,720	2,010	1,760	1,700	1,880
	(MJ)	8.8	6.8	8.8	7.3	6.8	6.7	7.2	8.4	7.4	7.1	7.9
Total protein	(g)	71.0	54.1	72.4	59.6	54.7	51.6	55.0	69.3	58.9	52.8	61.2
Animal protein	(g)	43.8	32.9	45.3	36.5	33.3	30.6	31.7	43.5	36.1	30.8	35.7
Fat	(g)	97	75	98	79	76	73	79	95	82	76	86
Fatty acids:												
saturated	(g)	38.8	28.5	38.5	31.2	29.5	28.1	29.7	37.6	31.5	27.9	32.2
monounsaturated	(g)	35.5	27.8	36.0	29.2	28.0	27.1	29.2	34.8	30.8	28.2	32.0
polyunsaturated	(g)	15.3	13.0	15.9	12.9	13.0	12.8	14.6	15.4	14.3	14.6	15.4
Carbohydrate	(g)	254	199	250	209	194	194	209	233	209	213	229
of which:												
non-milk extrinsic sugars	(g)	67	46	65	48	42	42	46	58	49	49	51
Calcium	(mg)	950	740	910	800	750	730	750	870	770	750	780
Iron	(mg)	11.5	8.8	11.4	9.6	9.1	8.5	9.3	10.8	9.3	8.4	10.0
Zinc	(mg)	9.0	6.8	9.1	7.5	6.9	6.4	6.9	8.7	7.3	6.5	7.6
Magnesium	(mg)	263	185	256	209	190	182	199	243	201	187	212
Sodium	(g)	2.83	2.28	2.80	2.43	2.26	2.17	2.36	2.70	2.38	2.00	2.42
Potassium	(g)	2.85	2.14	2.87	2.36	2.16	2.02	2.25	2.73	2.32	2.11	2.43
Thiamin	(mg)	1.44	1.10	1.43	1.19	1.10	1.06	1.18	1.35	1.18	1.10	1.26
Riboflavin	(mg)	1.88	1.43	1.81	1.54	1.43	1.38	1.46	1.71	1.48	1.36	1.52
Niacin equivalent	(mg)	27.8	21.7	28.8	23.5	21.7	20.4	22.0	27.5	23.6	20.9	24.3
Vitamin B6	(mg)	1.9	1.5	1.9	1.6	1.5	1.4	1.6	1.8	1.6	1.5	1.7
Vitamin B12	(µg)	5.8	4.2	5.8	4.4	4.0	3.7	3.8	5.5	4.6	4.2	4.9
Folate	(µg)	280	208	282	228	211	196	221	262	227	209	238
Vitamin C	(mg)	55	38	61	49	44	38	46	54	46	36	46
Vitamin A:												
retinol	(µg)	1,130	690	1,040	770	630	540	580	1,030	880	800	1,020
β-carotene	(µg)	2,020	1,390	2,100	1,650	1,510	1,290	1,620	2,000	1,600	1,400	1,390
total (retinol equivalent)	(µg)	1,470	920	1,390	1,050	880	750	850	1,370	1,150	1,040	1,260
Vitamin D	(µg)	3.82	2.41	3.6	2.65	2.51	2.38	2.80	3.06	2.54	2.27	2.66

(ii) As a percentage of Reference Nutrient Intake[a]

Energy[b]	98	91	95	86	82	80	89	90	83	86	85
Protein	141	155	141	141	139	135	152	136	129	132	123
Calcium	132	117	126	120	115	111	116	121	108	112	112
Iron	116	82	111	89	85	80	92	102	84	80	94
Zinc	110	98	107	99	91	84	91	103	93	87	93
Magnesium	91	86	88	85	81	78	90	84	75	78	75
Sodium	172	183	170	176	170	164	188	166	160	149	154
Potassium	79	85	80	81	80	76	90	77	73	76	71
Thiamin	164	152	160	147	139	134	153	151	139	140	143
Riboflavin	156	149	147	142	136	130	142	140	131	129	128
Niacin equivalent	195	182	195	176	165	154	172	184	167	160	166
Vitamin B6	147	147	145	140	135	128	146	138	130	134	131
Vitamin B12	377	372	376	344	332	313	333	365	336	339	330
Folate	136	131	137	129	125	116	138	129	121	123	121
Vitamin C	147	111	146	131	123	107	132	134	121	102	116
Vitamin A (retinol equivalent)	223	171	208	175	152	130	149	206	187	180	195

(iii) As a percentage of food energy

Fat	41.3	41.1	41.7	41.0	41.9	41.3	41.5	42.5	42.2	40.3	41.2
of which:											
saturated fatty acids	16.6	15.7	16.5	16.2	16.3	15.9	15.6	16.9	16.1	14.8	15.4
Carbohydrate	45.2	45.6	44.5	45.2	44.6	45.7	45.7	43.7	44.5	47.2	45.8

(iv) Contributions to selected nutrients from soft and alcoholic drinks and confectionery

Energy	(kcal)	100	100	100	110	110	110	110	90	90	80	70
	(MJ)	0.4	0.4	0.4	0.4	0.4	0.5	0.4	0.4	0.4	0.3	0.3
Fat	(g)	1	1	1	1	1	2	1	1	1	1	1
Carbohydrate	(g)	15	21	15	18	21	22	23	15	18	17	13

(a) Department of Health, *Dietary Reference Values for Food Energy and Nutrients for the United Kingdom,* HMSO, 1991
(b) as a percentage of Estimated Average Requirement

Table B13
Contributions made by selected foods to the nutritional value of household food: national averages, 1992

per person per day

	Energy	Fat	Fatty Acids Saturated	Fatty Acids Poly-unsaturated	Total Sugars[a]	Starch[b]	Fibre[c]
	kcal	g	g	g	g	g	g
Milk and milk products	196	9.4	5.9	0.3	18.1	0.1	...
of which: whole milk	95	5.7	3.6	0.2	6.8	−	−
low fat milks	62	1.8	1.1	0.1	7.1	−	−
yoghurt	15	0.2	0.1	...	2.6	0.1	−
Cheese	60	5.0	3.1	0.2	0.1	−	−
Meat and meat products	283	21.4	8.3	2.1	0.6	4.4	0.2
of which: carcase meat	98	8.0	3.2	0.6	−	−	−
poultry, uncooked	34	2.2	0.7	0.4	−	−	−
bacon and ham	36	3.0	1.1	0.3	...	−	−
offal	...	−	−	−	−	−	−
Fish	29	1.6	0.4	0.5	0.1	0.7	...
Eggs	22	1.6	0.5	0.2	−	−	−
Fats	248	27.2	8.8	6.5	0.4	...	−
of which: butter	43	4.8	3.2	0.2	−	−	−
margarine	83	9.2	2.4	2.9	0.1	−	−
low fat and dairy spreads	37	4.0	1.3	0.3	0.1	−	−
vegetable and salad oils	58	6.4	0.7	2.6	−	−	−
Sugar and preserves	105	...	...	...	27.9	0.1	...
Vegetables	192	4.3	1.1	1.5	6.8	27.5	4.8
of which: fresh potatoes	74	0.1	...	0.1	1.1	16.1	1.1
fresh green vegetables	6	0.2	...	0.1	0.6	0.1	0.4
other fresh vegetables	15	0.2	...	0.1	2.6	0.2	0.9
frozen vegetables	24	0.6	0.1	0.2	0.6	3.4	0.8
canned vegetables	20	0.2	...	0.1	1.4	2.3	0.9
Fruit	72	1.4	0.4	0.4	13.9	0.5	1.3
of which: fresh fruit	35	0.2	...	...	7.8	0.4	1.0
fruit juices	12	−	−	−	3.0	−	...
Cereals	599	12.1	4.7	2.0	18.5	94.7	5.5
of which: white bread (standard loaves)	117	0.8	0.1	0.2	1.5	23.3	0.8
brown and wholemeal bread	63	0.7	0.1	0.2	0.6	11.7	1.4
cakes, pastries and biscuits	158	6.8	3.2	0.7	9.8	13.4	0.7
breakfast cereals	66	0.5	0.1	0.2	3.6	11.0	1.2
Other foods	52	1.9	0.5	0.7	5.9	1.6	0.2
Total food	**1,859**	**86.0**	**33.6**	**14.4**	**92.2**	**129.6**	**12.0**
Soft drinks	44	−	−	−	11.8	−	−
Alcoholic drinks	24	...	−	−	0.7	−	−
Confectionery	32	1.2	0.7	0.1	4.4	0.5	−
Total food and drink	**1,960**	**87.2**	**34.3**	**14.5**	**109.1**	**130.1**	**12.0**

(a) includes sucrose, glucose, fructose, lactose and other simple sugars, as their monosaccharide equivalents
(b) as its monosaccharide equivalent
(c) as non-starch polysaccharides

Table B13 continued

per person per day

	Calcium	Iron	Sodium[d]	Vitamin C	Vitamin A[e]	Vitamin D
	mg	mg	mg	mg	µg	µg
Milk and milk products	384	0.2	181	2.6	132	0.21
of which: whole milk	162	0.1	78	0.9	77	0.04
low fat milks	170	0.1	78	0.9	25	0.01
yoghurt	27	...	12	0.2	3	...
Cheese	99	...	112	–	53	0.04
Meat and meat products	22	1.8	493	0.9	440	0.02
of which: carcase meat	3	0.6	23	–	2	–
poultry, uncooked	1	0.1	13	–	6	–
bacon and ham	1	0.1	185	0.1	–	–
offal	...	...	1	–	–	–
Fish	15	0.2	54	...	2	0.65
Eggs	9	0.3	21	–	29	0.26
Fats	5	...	177	–	223	1.45
of which: butter	1	...	44	–	52	0.04
margarine	2	...	80	–	99	0.90
low fat and dairy spreads	1	–	52	0.1	70	0.50
vegetable and salad oils	–	–	–	–	–	–
Sugar and preserves	3	0.1	3	0.5	0.2	–
Vegetables	49	1.7	249	21.5	237	...
of which: fresh potatoes	5	0.4	10	6.7	–	–
fresh green vegetables	9	0.2	1	2.6	10	–
other fresh vegetables	12	0.2	9	4.9	172	–
frozen vegetables	6	0.3	11	3.7	31	–
canned vegetables	11	0.4	120	0.9	16	–
Fruit	18	0.3	12	23.9	6	–
of which: fresh fruit	10	0.1	3	12.1	5	–
fruit juices	3	0.1	3	10.8	–	–
Cereals	200	4.8	947	0.7	16	0.30
of which: white bread	54	0.8	276	–	–	–
brown and wholemeal bread	23	0.8	160	–	–	–
cakes, pastries and biscuits	34	0.7	129	...	6	0.03
breakfast cereals	8	1.4	121	0.4	–	0.23
Other foods	23	0.5	264	0.7	13	0.02
Total food	**826**	**10.1**	**2,513**	**50.7**	**1,154**	**2.97**
Soft drinks	4	...	11	3.8	–	–
Alcoholic drinks	3	0.1	3	–	–	–
Confectionery	10	0.1	6	–	–	–
Total food and drink	**843**	**10.3**	**2,532**	**54.4**	**1,154**	**2.97**

(d) excludes sodium from table salt
(e) retinol equivalent

Glossary

Glossary of terms used in the Survey

Adult   A person of 18 years of age or over; however, solely for purposes of classifying households according to their composition, heads of household and diary-keepers under 18 years of age are regarded as adults.

Average consumption   The aggregate amount of food obtained for consumption (q.v.) by the households in the sample divided by the total number of persons in the sample.

Average expenditure   The aggregate amount spent by the households in the sample divided by the total number of persons in the sample.

Average price   Sometimes referred to as 'average unit value'. The aggregate expenditure by the households in the sample on an item in the Survey Classification of foods, divided by the aggregate quantity of that item purchased by these households.

Child   A person under 18 years of age; however, solely for purposes of classifying households according to their composition, heads of household and diary-keepers under 18 years of age are regarded as adults.

Convenience foods   Those processed foods for which the degree of preparation has been carried to an advanced stage by the manufacturer and which may be used as labour-saving alternatives to less highly processed products. The convenience foods distinguished by the Survey are cooked and canned meats, meat products (other than uncooked sausages), cooked and canned fish, fish products, canned vegetables, vegetable products, canned fruit, fruit juices, cakes and pastries, biscuits, breakfast cereals, instant coffee and coffee essences, baby foods, canned soups, dehydrated soups, ice-cream, and all frozen foods which fulfil the requirements of the previous sentence.

Food obtained for consumption   Food purchases from all sources (including purchases in bulk) made by households during their week of participation in the Survey and intended for human consumption during that week or later, plus any garden or allotment produce, etc (q.v.) which households actually consumed while participating in the Survey, but excluding sweets, alcohol, soft drinks and meals or snacks purchased to eat outside the home. For an individual household, the quantity of food thus obtained for consumption, or estimates of nutrient intake derived from it, may differ from actual consumption because of changes in household stocks during the week and because of wastage. Averaged over a sufficiently large group of households and a sufficiently long period of time, increases in household stocks might reasonably be expected to differ but little from depletions provided other things remain equal.

Garden and allotment produce, etc Food which enters the household without payment, for consumption during the week of participation in the Survey; it includes supplies obtained from a garden, allotment or farm, or from an employer, but not gifts of food from one household in Great Britain to another if such food has been purchased by the donating household. (See also 'Value of garden and allotment produced, etc').

Household For the Survey purposes, this is defined as a group of persons living in the same dwelling and sharing common catering arrangements.

Income group Households are grouped into eight income groups (A1, A2, B, C, D, E1, E2 and OAP) according to the ascertained or estimated gross income of the head of the household, or of the principal earner in the household if the weekly income of the head is less than the amount defining the upper limit to income group D.

Intake See 'Food obtained for consumption'

Net balance The net balance of an individual (a member of the household or a visitor) is a measure of the proportion of the individuals' food needs met by meals eaten in the home by that individual during the Survey week, each meal being given a weight in proportion to its importance. The relative weights currently used are breakfast 3, mid-day meal 4, evening meal 7. These weights were changed during 1991; previously, separate weights for tea (2) and supper (5) were used if two evening meals were taken; now a light tea or supper is disregarded in this calculation. The net balance is used when relating nutrient intake to need.

Nutrients In addition to the energy value of food expressed in terms of kilocalories and megajoules (4,184 megajoules = 1,000 kilocalories), the food is evaluated in terms of the following nutrients:

Protein (animal and total), fat (including the component saturated, monounsaturated and polyunsaturated fatty acids), carbohydrate, calcium, iron, zinc, magnesium, sodium, potassium, vitamin A (retinol, β-carotene, retinol equivalent), thiamin, riboflavin, niacin equivalent, folate, vitamins B6, B12, C and D.

Pensioner households (OAP) Households in which at least three-quarters of total income is derived from state retirement pensions or similar pensions and/or supplementary pensions or allowances paid in supplementation or instead of such pensions. Such households will include at least one person over the state retirement age.

Person An individual of any age who during the week of the Survey spends at least four nights in the household ('at home') and has at least one meal a day from the household food supply on at least four days, except that if he/she is the head of the household, or the diary-keeper, he or she is regarded as a person in all cases.

Price index A price index of Fisher 'Ideal' type is used; this index is the geometric mean of two indices with weights relating to the earlier and later periods respectively, or in the case of non-temporal comparisons (eg regional, type of area, income group and household composition), with weights relating to the group under consideration and the national average respectively.

Quantity index This is also an index of Fisher 'Ideal' type. The price and quantity indices together account for the whole of the expenditure difference between the two periods or groups being compared.

Real price The price of an item of food in relation to the price of all goods and services. The term is used when refering to changes in the price of an item over a period of time. It is measured by dividing the average price (q.v.) paid at a point in time by the General Index of Retail Prices (All Items) at that time.

Regions The standard regions for statistical purposes, except that East Anglia is combined with the South East region.

Seasonal foods Those foods which regularly exhibit a marked seasonal variation in price or in consumption; for the purposes of the Survey these are deemed to be eggs, fresh and processed fish, shellfish, potatoes, fresh vegetables and fresh fruit.

Value of consumption Expenditure plus value of garden and allotment produce, etc (q.v.)

Value of garden and allotment produce, etc The value imputed to such supplies received by a group of households is derived from the average prices currently paid by the group for corresponding purchases. This appears to be the only practicable method of valuing these supplies, though if the households concerned had not had access to them, they would probably not have consumed as much of these foods, and would therefore have spent less on them than the estimated value of their consumption. Free school milk and free welfare milk are valued at the average price paid by the group for full price milk. (See also 'Garden and allotment produce, etc').

Symbols and conventions used

Symbols The following are used throughout:

    − = nil

    ... = less than half the final digit shown

na or blank = not available or not applicable

Rounding of figures In tables where figures have been rounded to the nearest final digit, there may be an apparent slight discrepancy between the sum of the constituent items and the total shown.

Additional Information

Analyses of Survey data providing more detail and, in some cases, more-up-to-date information than published in this report are available directly from the Ministry of Agriculture, Fisheries and Food. These analyses are of two main types:

i) Standard analyses
    Quarterly national averages - available approximately 10 weeks after the end of each survey period
    Analyses of components of selected food codes

ii) Ad hoc analyses
    Ad hoc analyses can be undertaken to meet the special requirements of organisations, subject to resources being available

Further details regarding additional Survey information are available from:

National Food Survey Branch
Ministry of Agriculture, Fisheries and Food
Room 513, West Block
Whitehall Place
London SW1A 2HH

Telephone: 071-270-8562/3